Energy Efficient Cooperative Wireless Communication and Networks

OTHER COMMUNICATIONS BOOKS

Anonymous Communication Networks: Protecting Privacy on the Web
Kun Peng
ISBN 978-1-4398-8157-6

Case Studies in System of Systems, Enterprise Systems, and Complex Systems Engineering
Alex Gorod, Brian E. White, Vernon Ireland, S. Jimmy Gandhi, and Brian Sauser (Editors)
ISBN 978-1-4665-0239-0

Cyber-Physical Systems: Integrated Computing and Engineering Design
Fei Hu
ISBN 978-1-4665-7700-8

Evolutionary Dynamics of Complex Communications Networks
Vasileios Karyotis, Eleni Stai, and Symeon Papavassiliou
ISBN 978-1-4665-1840-7

Fading and Interference Mitigation in Wireless Communications
Stefan Panic, Mihajlo Stefanovic, Jelena Anastasov, and Petar Spalevic
ISBN 978-1-4665-0841-5

Green Networking and Communications: ICT for Sustainability
Shafiullah Khan and Jaime Lloret Mauri (Editors)
ISBN 978-1-4665-6874-7

Intrusion Detection in Wireless Ad-Hoc Networks
Nabendu Chaki and Rituparna Chaki (Editors)
ISBN 978-1-4665-1565-9

Intrusion Detection Networks: A Key to Collaborative Security
Carol Fung and Raouf Boutaba
ISBN 978-1-4665-6412-1

Machine-to-Machine Communications: Architectures, Technology, Standards, and Applications
Vojislav B. Mišić and Jelena Mišić (Editors)
ISBN 978-1-4665-6123-6

MIMO Processing for 4G and Beyond: Fundamentals and Evolution
Mário Marques da Silva and Francisco A. Monteiro (Editors)
ISBN 978-1-4665-9807-2

Network Innovation through OpenFlow and SDN: Principles and Design
Fei Hu (Editor)
ISBN 978-1-4665-7209-6

Opportunistic Mobile Social Networks
Jie Wu and Yunsheng Wang (Editors)
ISBN 978-1-4665-9494-4

Physical Layer Security in Wireless Communications
Xiangyun Zhou, Lingyang Song, and Yan Zhang (Editors)
ISBN 978-1-4665-6700-9

SC-FDMA for Mobile Communications
Fathi E. Abd El-Samie, Faisal S. Al-kamali, Azzam Y. Al-nahari, and Moawad I. Dessouky
ISBN 978-1-4665-1071-5

Security for Multihop Wireless Networks
Shafiullah Khan and Jaime Lloret Mauri (Editors)
ISBN 978-1-4665-7803-6

Self-Healing Systems and Wireless Networks Management
Junaid Ahsenali Chaudhry
ISBN 978-1-4665-5648-5

The State of the Art in Intrusion Prevention and Detection
Al-Sakib Khan Pathan (Editor)
ISBN 978-1-4822-0351-6

Wi-Fi Enabled Healthcare
Ali Youssef, Douglas McDonald II, Jon Linton, Bob Zemke, and Aaron Earle
ISBN 978-1-4665-6040-6

Wireless Ad Hoc and Sensor Networks: Management, Performance, and Applications
Jing (Selina) He, Shouling Ji, Yingshu Li, and Yi Pan
ISBN 978-1-4665-5694-2

Wireless Sensor Networks: From Theory to Applications
Ibrahiem M. M. El Emary and S. Ramakrishnan (Editors)
ISBN 978-1-4665-1810-0

ZigBee® Network Protocols and Applications
Chonggang Wang, Tao Jiang, and Qian Zhang (Editors)
ISBN 978-1-4398-1601-1

Energy Efficient Cooperative Wireless Communication and Networks

Edited by

Zhengguo Sheng • Chi Harold Liu

CRC Press
Taylor & Francis Group
Boca Raton London New York

CRC Press is an imprint of the
Taylor & Francis Group, an **informa** business

CRC Press
Taylor & Francis Group
6000 Broken Sound Parkway NW, Suite 300
Boca Raton, FL 33487-2742

Printed on acid-free paper
Version Date: 20140623

International Standard Book Number-13: 978-1-4822-3821-1 (Hardback)

Library of Congress Cataloging-in-Publication Data

Energy efficient cooperative wireless communication and networks / Zhengguo Sheng and Chi Harold Liu, editors.
pages cm
Includes bibliographical references and index.
ISBN 978-1-4822-3821-1 (hardback)
1. Wireless communication systems--Energy conservation. 2. Wireless communication systems--Energy consumption. 3. Engineering economy. I. Liu, Chi Harold. II. Sheng, Zhengguo.

TK5102.86.E543 2014
621.384--dc23 2014019422

Visit the Taylor & Francis Web site at
http://www.taylorandfrancis.com

and the CRC Press Web site at
http://www.crcpress.com

Contents

Editors

Zhengguo Sheng is a lecturer at the University of Sussex, UK, and co-founder of WRTnode. His current research interests cover Internet-of-things (IoT), machine-to-machine (M2M) technologies, mobile cloud computing, and power line communication (PLC). Previously, he was with the University of British Columbia as a research associate and with France Telecom Orange Labs as the senior researcher and project manager in M2M and Internet-of-Things, as well as the coordinator of Orange and Asia telco on NFC-SWP partnership. He is also the winner of the Orange Outstanding Researcher Award and CEO Retention bonus recipient, 2012. He also worked as a research intern with IBM T. J. Watson Research Center, USA, and U.S. Army Research Labs. With six years of research experience across industry and academia, Sheng has research interests that cover a wide range in wireless communication from the fundamental information theory to radio technology and protocol design, and so on. Before joining Orange Labs, he received his Ph.D. and M.S. degrees with distinction at Imperial College London in 2011 and 2007, respectively, and his B.Sc. degree from the University of Electronic Science and Technology of China (UESTC) in 2006.

He has published more than 30 prestigious conference and journal papers. He serves as the technical committee member of Elsevier Journal of Computer Communications (COMCOM). He has also served as the co-organizer of IEEE International Symposium on Wireless Vehicular Communications (WiVeC'14), session chair of IEEE VTC'14–Fall, and technical program committee members of Tensymp'15, CloudCom'14, SmartComp'14, WCSP'14, Qshine'14, ICCAAD'14, ContextDD'14, and others. He is also a member of the Institute of Electrical and Electronics Engineers (IEEE), Vehicular Technology Society (VTS), and the Association for Computing Machinery (ACM).

Chi Harold Liu is a full professor at the School of Software, Beijing Institute of Technology, China. He is also the director of the IBM Mainframe Excellence Center (Beijing), director of the IBM Big Data and Analysis Technology Center, and director of the National Laboratory of Data Intelligence for China Light Industry. He holds a Ph.D. degree from Imperial College, UK, and a B.Eng. degree from Tsinghua University, China. Before moving to academia, he joined IBM Research–China as a staff researcher and project manager, and worked as a postdoctoral researcher at Deutsche Telekom Laboratories, Germany, and as a visiting scholar at IBM T. J. Watson Research Center, USA. His current research interests include the Internet-of-Things (IoT), big data analytics, mobile computing, and wireless ad hoc, sensor, and mesh networks. He received the Distinguished Young Scholar Award in 2013, IBM First Plateau Invention Achievement Award in 2012, and IBM First Patent Application Award in 2011 and was interviewed by EEWeb.com as the Featured Engineer in 2011. He has published more than 50 prestigious conference and journal papers and owned more than 10 EU/U.S./China patents. He serves as the editor for *KSII Transactions on Internet and Information Systems* and the book editor for four books published by Taylor & Francis Group, USA. He also has served as the general chair of IEEE SECON'13 workshop on IoT Networking and Control, IEEE WCNC'12 workshop on IoT Enabling Technologies, and ACM UbiComp'11 Workshop on Networking and Object Memories for IoT. He served as a consultant to Bain & Company and KPMG, USA, and as a peer reviewer for the Qatar National Research Foundation and National Science Foundation, China. He is a member of IEEE and ACM.

Acknowledgments

The editors would like to thank all the chapter authors for their invaluable contributions to this book and Prof. Victor C.M. Leung, from The University of British Columbia, and Prof. Kin K. Leung, from Imperial College London, for their helpful discussions and comments that improved the quality of the book.

This book is supported by the Canadian Natural Sciences and Engineering Research (NSERC), the NSERC DIVA Strategic Research Network, and various industry partners.

The editors would also like to thank National Natural Science Foundation of China (Grant No: 61300179).

CHAPTER 1

Introduction

Zhengguo Sheng, University of British Columbia, Canada
Chi Harold Liu, Beijing Institute of Technology, China

1.1 OVERVIEW

In the last few years, there has been a lot of interest in wireless ad hoc networks as they have remarkable commercial and military applications. Such wireless networks have the benefit of avoiding a wired infrastructure. However, signal fading is a severe problem for wireless communications, particularly for the multi-hop transmissions in the ad hoc networks. In real application, without considering this issue, signals may not be received properly. In order to deal with this problem, the use of diversity provides a good way to reduce signal interference. The multiple-input-multiple-output (MIMO) antenna can provide spatial diversity and multiplexing gain in wireless networks. It also represents a powerful technique for interference mitigation and reduction. Therefore, to meet the needs of future wireless communications, it is advantageous to equip the associated wireless ad hoc networks with MIMO antenna capabilities.

Toward this goal, cooperative communication is studied as an alternative and low-cost way to achieve spatial diversity. The key feature of cooperative transmission is to encourage multiple single-antenna users/sensors to share their antennas cooperatively. In this way, a virtual antenna array can be constructed, and as a result, the overall quality of the wireless transmission, in terms of the reception reliability [1, 2], energy efficiency [3], and network capacity [4], can be improved significantly. Cooperative diversity has largely been considered by physical layer researchers, and various cooperative transmission protocols have been developed at the physical layer to further increase the bandwidth efficiency of cooperative diversity. There have been intensive studies on the physical layer techniques of cooperative communication; we refer the interested reader to some state-of-the-art works [1, 2] for a preliminary understanding of cooperative transmission at the physical layer.

Moreover, the benefits of cooperative communication can be further exploited by the significant interactions between various layers of the protocol stack for performance enhancements. Opportunistic scheduling is a good example of cross-layer design, where scheduling protocols are designed by taking advantage of the knowledge of wireless link conditions [5, 6]. It has been shown that these cross-layer designs [7] and protocols could be essential for wireless ad hoc and sensor networks where unpredictable variables such as node mobility, node density, and network dimensions make the diverse and stringent wireless quality-of-service (QoS) requirements difficult to satisfy.

On the way to the development of cooperative communication, there is a considerable need to understand its practical benefits and limitations, and its interdependence with networking functions. Especially, it becomes critically important to study how the performance gain of cooperative diversity at the physical layer can be reflected at the network layer, thus ultimately improving application performance.

1.2 RELATED WORK

More recent works in the literature show that cooperative communication can significantly improve the overall quality of the wireless transmission. Lee et al. [8] examined the symbol-error-rate (SER) performance of decode-and-forward (DF) cooperative communications with multiple dual-hop relays over Nakagami-m fading channels and showed that SER performance is significantly improved with channel conditions or fading parameters, because of the increased diversity order. Asghari and Aissa [9] also considered that in spectrum-sharing systems the error performance of cognitive (secondary) users' communication can be significantly improved by implementing the partial relay selection using DF without affecting the performance of licensed (primary) users. Meanwhile, Tao and Liu [10] addressed an optimization problem involving transmission mode selection (direct or cooperative transmission), relay selection, and subcarrier assignment to maximize throughput in cooperative OFDMA networks. Their work showed that the proposed algorithm can enhance throughput performance by more than 75% compared to direct transmission. Elhawary and Haas [11] proposed an energy-efficient routing protocol for cooperative networks by employing relay clusters along a non-cooperative path and revealed that the proposed cooperative transmission protocol can save up to 40% of energy compared with the disjoint-paths and the one-path scheme using only direct transmission.

There are also existing works on the analysis of delay and network capacity of wireless networks by using the concept of cooperation. Xu et al. [12] measured queuing delay in a two-user cooperation system, and the proposed scheduling policy is proven to greatly reduce the delay imbalance between users. Also Song et al. [13] showed that the connectivity of wireless networks can be significantly improved through collaboration. Furthermore, the collaborative networks require less power than noncollaborative networks in order to maintain connectivity of the whole network. Although various cooperative transmission schemes have been developed to increase the bandwidth efficiency, most existing literature focuses on physical layer techniques, and there is a lack of understanding of cooperative benefits at the upper layers (e.g., routing and applications).

In the latest work of relay selection, Babaee and Beaulieu [14] proposed an optimization problem to find a number of relays along the path with the minimum end-to-end outage probability from source to destination. The proposed solution requires an optimization over all the paths connecting source-destination subject to a fixed total power constraint. Ikki and Ahmed [15] investigated the performance of the best-relay selection and showed that the best-relay selection not only reduces the amount of required resources but also can maintain a full diversity order. Cho et al. [16] also proposed a best relay selection scheme to ensure minimum outage probability given a Poisson field of relay nodes and the presence of path loss and fading, and argued that relays geographically approaching the source and destination are preferred to others.

1.3 MOTIVATION AND AIMS

Although much work related to cooperative communication has been carried out since the early 2000s, it is not adequate for the following reasons. First, to the best of our knowledge, existing work typically considers a single transmitter serving one or multiple users. Mutual interference, interdependency, and dynamics among multiple transmitting nodes in ad hoc networks using cooperative transmission have not been considered in the protocol/control design from the cross-layer perspective. Second, how the cooperative transmission and its associated protocols in the multi-transmitter scenarios affect upper layers' performance is not well understood. Evidently, the optimal system and protocol design represents a very complicated problem and is still open.

The goal of this book is to study cooperative transmission and the associated designs of upper-layer protocols, including MAC, routing, and transport protocol, and ultimately to improve the overall QoS at the application level in the wireless networks. Therefore, there is a considerable need to understand its practical benefits and limitations, and its interdependence with networking functions. Especially, it becomes critically important to study how the performance gain of cooperative diversity at the physical layer can be reflected at the network layer, thus ultimately improving application performance. Different from the above literature, our contribution in this book is that we consider power efficiency as a main object in cooperative communication to ensure QoS requirements. To be specific, it is of fundamental importance to understand (1) how to bring the performance gain at the physical layer up to the network layer and (2) how to allocate network resources dynamically through MAC/scheduling and routing so as to trade off the performance benefit of a given transmission (optimized by allocating many cooperating nodes) against network cost (power, interference, coordination overhead, and delay). The selected techniques in each chapter can help achieve the global energy efficiency as well as reliability in wireless networks. Hence these results will potentially have a broad impact across a range of industry areas, including wireless communication, wireless sensor/ad hoc networks, and so on.

1.4 ORGANIZATION OF THE BOOK

We focus on the development of performance effective algorithms for cooperative wireless networks and various aspects of their system performance. Our starting point is to provide a fundamental understanding of the physical layer technique, which lays a foundation to develop network protocols for practical environments. With a better understanding of the cooperative transmission mechanism, it becomes critically important to examine how the performance gain of cooperative diversity at the physical layer can be reflected at the networking layer. Our approach is to achieve that by tailoring the designs of network protocols for cooperative communication and further evaluating its network performance. Especially, we investigate end-to-end performance (i.e., reliability, throughput, power, and delay) of a multi-hop cooperative route by introducing new techniques (e.g., interference subtraction and supplementary cooperation). Finally, the whole analysis of cooperative networks is extended to a more general network scenario where multi-pair multi-transmission coexists.

I

Fundamental Understanding of Cooperative Communication

CHAPTER 2

Reliability of Cooperative Transmission

Zhiguo Ding, Newcastle University, UK
Zhengguo Sheng, University of British Columbia, Canada

2.1 SYSTEM MODEL

We start with a direct transmission link as depicted in Figure 2.1(a) and assume the channel model incorporating path-loss and Rayleigh fading. The received signal at the destination d is modeled as

$$y_d[n] = \mathrm{a}_{s,d} x_s[n] + n_d[n], \tag{2.1}$$

where $x_s[n]$ is the signal transmitted by a source s, $n \in [1, ..., N]$ is the index of the transmitting packet, and $n_d[n]$ is additive white Gaussian noise, with variance σ_n^2, at the receiver. The channel gain $\mathrm{a}_{s,d}$ between the nodes s and d is modeled as $\mathrm{a}_{s,d} = h_{s,d}/d_{s,d}^{\alpha/2}$, where $d_{s,d}$ is the distance between the nodes s and d, α is the path-loss exponent, and $h_{s,d}$ captures the channel fading characteristics. The channel fading parameter $h_{s,d}$ is assumed to be complex Gaussian with zero mean and unit variance, and independent and identically distributed (i.i.d.) across times slots, packets, and links.

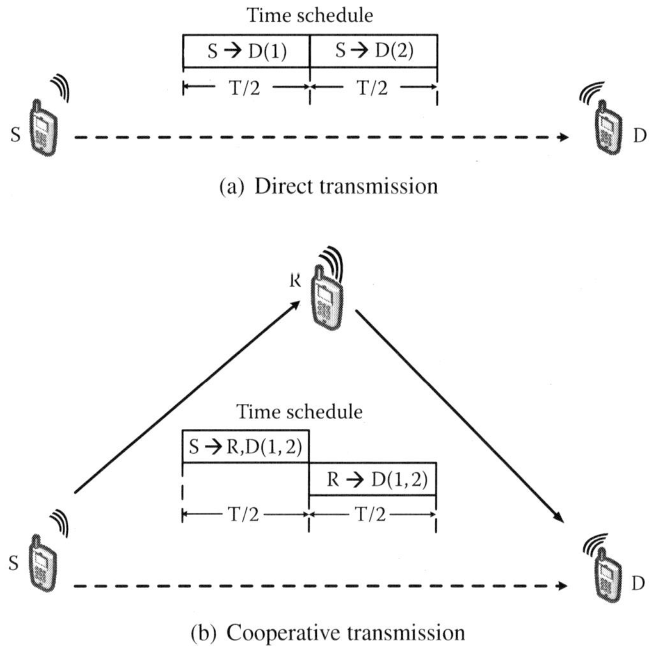

FIGURE 2.1 Comparison of direct transmission and cooperative transmission.

In this chapter, we consider the selected decode-and-forward (DAF) cooperative scheme [1] in our system model, since this cooperative scheme lends itself to a relatively easy implementation in hardware and software. The scenario is depicted in Figure 2.1(b), where a source and destination communicate to each other with help of one single relay. Each node is equipped with one omnidirectional antenna. Here, relay transmission is a main feature of cooperative communication.

As defined in this book, a cooperative link (CL) between the source and destination nodes includes two different transmission channels. The dashed line is the direct transmission channel from the source directly to the destination, while the combined solid lines are relay transmission channels from the source through the relay to the destination. In order to overcome the inability of current radio frequency (RF) capture effects when simultaneously transmitting and receiving in the same frequency band, the communication is divided into two orthogonal time slots:

- **In the first time slot:** the source broadcasts its data to the relay and the destination and they receive

$$\begin{aligned} y_r[n] &= \frac{h_{s,r}}{d_{s,r}^{\alpha/2}} x_s[n] + n_r[n], \\ y_{d,1}[n] &= \frac{h_{s,d}}{d_{s,d}^{\alpha/2}} x_s[n] + n_d[n], \end{aligned} \tag{2.2}$$

  where $d_{s,d}$, $d_{s,r}$, and $d_{r,d}$ are the respective distances among the source, relay, and destination node, $x_s[n]$ is the information transmitted by the source, and $n_d[n]$ and $n_r[n]$ are white noise.

- **In the second time slot:** the relay transmits the signal it received in the previous time slot, if it can decode the signal successfully (i.e., the received SNR exceeds a threshold); otherwise, the source retransmits the signal to the destination. Thus an ACK from relay to source is assumed. The destination node receives

$$y_{d,2}[n] = \begin{cases} \dfrac{h_{s,d}}{d_{s,d}^{\alpha/2}} x_s[n] + n_d[n], & \text{if SNR}_{s,r} < \eta, \\[2ex] \dfrac{h_{r,d}}{d_{r,d}^{\alpha/2}} x_r[n] + n_d[n], & \text{if SNR}_{s,r} \geq \eta, \end{cases} \tag{2.3}$$

where η is a threshold value to guarantee a successful decoding at the relay node.

As a result, the destination receives two independent copies of the same packets transmitted through different wireless channels. Diversity gain can be achieved by combining the data copies using one of a variety of combining techniques, for example, the Maximum Ratio Combining (MRC) [17] where the received signals are weighted with respect to their SNR and then summed together. A full second order of diversity can be obtained from such a cooperative transmission strategy [1]. Such cooperative communication brings significant improvement of reception reliability, which becomes an important criterion to measure the performance of cooperative transmissions.

It is worth noting that the Time Division Multiple Access (TDMA) scheme is considered here for two reasons. First, in order to simplify the problem and in case there are multiple source-destination pairs communicating simultaneously, the TDMA assumption could allow us to only concentrate on one pair, and hence remove co-channel interference between the terminals at the destination automatically. Second, the fact that time division duplex channels are reciprocal naturally makes channel state information (CSI) available at the transmitter. To simplify the development of the proposed routing protocol, we consider that only one relay is used for cooperative transmission, whereas results for using multiple relays can be found [18].

2.2 OUTAGE BEHAVIOR OF TRANSMISSION SCHEMES

We measure the reception reliability in terms of outage probability, which is defined as follows.

2.2.1 Direct Transmission

We start with direct transmission, and the channel capacity between the source s and the destination d is

$$I_{s,d} = \log(1 + p|\mathrm{a}_{s,d}|^2), \tag{2.4}$$

where $p = \frac{E_b}{N_0}$ is defined as the normalized transmission power. For Rayleigh fading, $|\mathrm{a}_{s,d}|^2$ is exponentially distributed with parameter $d_{s,d}^{\alpha}$. The outage probability satisfies

$$\epsilon^{\text{out}} = \Pr[I_{s,d} < b] \;=\; 1 - \exp\left(-\frac{(2^b-1)d_{s,d}^{\alpha}}{p}\right)$$
$$\approx\; d_{s,d}^{\alpha}\left(\frac{2^b-1}{p}\right) \tag{2.5}$$

for large p. Here b is the desired data rate in bps/Hz, which is defined by QoS requirement. We then have the normalized transmission power for direct transmission

$$p_D = d_{s,d}^{\alpha}\left(\frac{2^b-1}{\epsilon^{\text{out}}}\right). \tag{2.6}$$

2.2.2 Cooperative Transmission

Let $d_{s,d}$, $d_{s,r}$, and $d_{r,d}$ be the respective distances among the source, relay, and destination node. During the first time slot, the destination and relay receive $y_{d,1}[n] = (h_{s,d}/d_{s,d}^{\alpha/2})x_s[n] + n_d[n]$ from the source node, where $x_s[n]$ is the information transmitted by the source and $n_d[n]$ is white noise. During the second time slot, the destination node receives

$$y_{d,2}[n] = \begin{cases} \dfrac{h_{s,d}}{d_{s,d}^{\alpha/2}}x_s[n] + n_d[n], & \text{if } \left|\dfrac{h_{s,r}}{d_{s,r}^{\alpha/2}}\right|^2 < f(p), \\[2ex] \dfrac{h_{r,d}}{d_{r,d}^{\alpha/2}}x_r[n] + n_d[n], & \text{if } \left|\dfrac{h_{s,r}}{d_{s,r}^{\alpha/2}}\right|^2 \geq f(p), \end{cases} \tag{2.7}$$

where $f(p) = \frac{2^{2b}-1}{p}$ can be derived from direct transmission and is analogous to (2.5). In this protocol, the relay transmits only if the SNR exceeds a threshold; otherwise, the source retransmits in the second time slot. We thus implicitly assume a mini-slot at the beginning of the second slot during which ACKs are sent error-free from the relay to the source.

Assuming that the relay node can perform perfect decoding when the received SNR exceeds a threshold, the channel capacity of this cooperative link can be shown as

$$I_{s,d} = \begin{cases} \frac{1}{2}\log(1+2p|\mathrm{a}_{s,d}|^2), & |\mathrm{a}_{s,r}|^2 < f(p), \\ \frac{1}{2}\log(1+p|\mathrm{a}_{s,d}|^2 + p|\mathrm{a}_{r,d}|^2), & |\mathrm{a}_{s,r}|^2 \geq f(p), \end{cases} \tag{2.8}$$

where p is the normalized transmission power for both source and relay. It is worth noting that the same noise variance is assumed at both relay and destination. Therefore, the outage event is given by $I_{s,d} < b$ and the outage probability becomes

$$\begin{aligned} \epsilon^{\text{out}} &= \Pr[I_{s,d} < b] \\ &= \Pr[|\mathrm{a}_{s,r}|^2 < f(p)]\Pr[2|\mathrm{a}_{s,d}|^2 < f(p)] \\ &\quad + \Pr[|\mathrm{a}_{s,r}|^2 \geq f(p)]\Pr[|\mathrm{a}_{s,d}|^2 + |\mathrm{a}_{r,d}|^2 < f(p)]\,. \end{aligned} \tag{2.9}$$

By computing the limit, we obtain from (2.9)

$$\begin{aligned} \frac{1}{f^2(p)}\epsilon^{\text{out}} &= \underbrace{\frac{1}{f(p)}\Pr[|\mathrm{a}_{s,r}|^2 < f(p)]}_{\textbf{T1}}\,\underbrace{\frac{1}{f(p)}\Pr[2|\mathrm{a}_{s,d}|^2 < f(p)]}_{\textbf{T2}} \\ &\quad + \underbrace{\Pr[|\mathrm{a}_{s,r}|^2 \geq f(p)]}_{\textbf{T3}}\,\underbrace{\frac{1}{f^2(p)}\Pr[|\mathrm{a}_{s,d}|^2 + |\mathrm{a}_{r,d}|^2 < f(p)]}_{\textbf{T4}}, \end{aligned} \tag{2.10}$$

where $\textbf{T1} \approx d_{s,r}^{\alpha}$, $\textbf{T2} \approx d_{s,d}^{\alpha}/2$, $\textbf{T3} \approx 1$, $\textbf{T4} \approx (d_{s,d}^{\alpha}d_{r,d}^{\alpha})/2$. Since $f(p) = \frac{2^{2b}-1}{p}$, we obtain a closed-form expression for the outage probability between the source and the destination using DAF cooperative transmission

$$\epsilon_C^{\text{out}} = \frac{1}{2}d_{s,d}^{\alpha}(d_{s,r}^{\alpha} + d_{r,d}^{\alpha})\frac{(2^{2b}-1)^2}{p^2}\,. \tag{2.11}$$

It is worth noting that for a fair comparison with direct transmission using only one time slot, cooperative transmission actually employs twice the data rate at $2b$; Figure 2.1 shows that the cooperative link transmits both packet 1 and 2 together, during two consecutive time slots, so that both schemes have the same effective data rate.

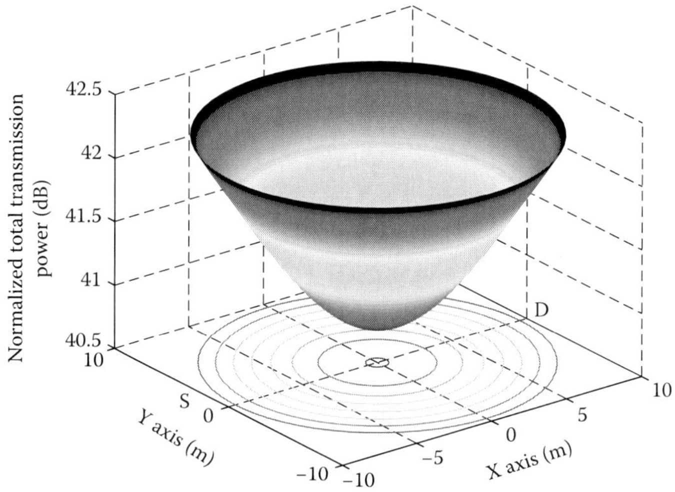

FIGURE 2.2 Total transmission power of cooperative transmission as a function of relay location.

Hence the total normalized power consumption for DAF cooperation is

$$p_{\mathrm{DAF}} = 2p = 2\sqrt{\frac{1}{2}d_{s,d}^{\alpha}(d_{s,r}^{\alpha} + d_{r,d}^{\alpha})\frac{(2^{2b}-1)^2}{\epsilon_C^{\mathrm{out}}}}. \tag{2.12}$$

2.3 MOTIVATING EXAMPLE

Readers may notice that the transmission power in (2.12) is not decided only by the data rate b and outage probability $\epsilon_C^{\mathrm{out}}$ but also by the relative distance between the relay and the source-destination pair.

To illustrate the main ideas, consider a three-node scenario where the relay node is located arbitrarily within the area defined by the circle whose diameter is the straight line between the source S and destination D. The distance between the nodes S and D is assumed to be $d_{s,d} = 20$ m, the required data rate is $b = 1$ bps/Hz, the prefixed outage probability is $\epsilon^{\mathrm{out}} = 0.01$, and the path-loss exponent is set as $\alpha = 2$. For such a system setup, the normalized power consumed by direct transmission is $p_D = 46$ dB, whereas the normalized power consumed by cooperative transmission is shown in Figure 2.2 as a function of relay location on the x-y plane. Recall that the nodes S and D are located at $(-10$ m, 0) and (10 m, 0), respectively; the cooperative transmission scheme can achieve its best performance if the relay node is located at the center of the circle. Moreover, the cooperative scheme always consumes less transmission power than the direct transmission.

It is clear that relay selection is crucial for the performance of cooperative transmission. This is so because a good-quality relay yields strong multiuser diversity gain, thus potentially enhancing the system performances (i.e., outage probability, transmission power, and data rate). Intuitively speaking, the larger reduction of power leads to less interference, thus conceivably increasing network capacity, reducing transmission delay, and so on, which is a main motivation to bring such performance gains up to the network layer. This will be discussed in the following chapters.

It is assumed that a route has been established between a source and a destination. Different from traditional routes, cooperative transmission is used to improve the link quality when the source node communicates with the destination node. The links involved in the route between the source and destination nodes can be categorized into two sets. The first set, defined as S_1, includes the links using direct transmission without using any relay, and the other one, defined as S_2, includes all links using cooperative transmission. In our model, we also assume identical transmission power for all nodes; thus the total transmission power is proportional to the total number of nodes involved in the route.

For the above scenario, by assuming that the error performances among links are independent, the end-to-end (ETE) outage probability can be derived from (2.5) and (2.11) and is given by

$$\epsilon_{\text{ETE}}^{\text{out}} = 1 - \prod_{i,j \in S_1} \left(1 - \epsilon_{i,j}^{\text{DT}}\right) \prod_{i,j \in S_2} \left(1 - \epsilon_{i,j}^{\text{CT}}\right), \tag{2.13}$$

where $\epsilon_{i,j}^{\text{DT}}$ and $\epsilon_{i,j}^{\text{CT}}$ denote outage probability for direct link and for cooperative link, respectively.

For small outage probability $\epsilon_{i,j}^{\text{DT}} \ll 1$ and $\epsilon_{i,j}^{\text{CT}} \ll 1$, we have the following approximation:

$$\epsilon_{\text{ETE}}^{\text{out}} \approx \sum_{i,j \in S_1} \epsilon_{i,j}^{\text{DT}} + \sum_{i,j \in S_2} \epsilon_{i,j}^{\text{CT}}. \tag{2.14}$$

Substituting (2.5) and (2.11) into the above, we obtain

$$\epsilon_{\text{ETE}}^{\text{out}} = \frac{(2^b - 1)}{p} \sum_{i,j \in S_1} d_{i,j}^{\alpha} + \frac{(2^{2b} - 1)^2}{2p^2} \sum_{i,j \in S_2} d_{i,j}^{\alpha} (d_{i,r}^{\alpha} + d_{r,j}^{\alpha}), \qquad (2.15)$$

where i, r, and j are source, relay, and destination nodes, respectively, of one cooperative link.

Based on different objective functions and constraints, even the same system setup could lead to different optimal routes [19]. Here we use (2.15) as our objective function to justify the end-to-end reliability of cooperative route.

2.4 DESCRIPTION OF THE QOS-DRIVEN ROUTING ALGORITHM

Based on the characteristics of cooperative transmission analyzed so far, we propose here a distributed routing algorithm to establish a cooperative route in an arbitrary network that ensures each link ϵ^{out} below a certain target level (constraint). Algorithm 2.1 describes the routing algorithm in detail.

The timing schedule of the proposed cooperative algorithm is shown in Figure 2.3. As a distributed routing algorithm, each relay node as a monitor periodically broadcasts a HELLO packet to its source-destination pair to measure the link performance. When an improvement is necessary, the relay sends a NOTIFICATION to its source and destination and triggers new relay selections among the source-relay and the relay-destination links. Such "control information" needs to be synchronized among the source, relay, and destination before packet transmission.

To fit the non-infrastructure nature of ad hoc networks, it is desirable to devise a distributed mechanism to choose the relay node with the best incoming and outgoing channel condition among candidate nodes without using a central controller. Specifically, the relay selection method is similar to the idea in Section 7.2.2, which employs a four-way handshake of messages to control medium accesses for cooperative communication.

In the proposed algorithm, relays use a similar carrier sensing scheme [20] and go through a backoff period before sending received data to the destination. Each relay then sets the backoff timer, proportional to its ϵ^{out}, and the node with the minimum backoff time shall (and is implicitly chosen to) relay the packet.

ALGORITHM 2.1 Proposed Cooperative Routing Algorithm

Input: Relay candidates $\mathcal{R}$, transmission power p, data rate b, outage constraint η
Output: Cooperative route R and link outage ϵ^{out}

Initialize:

$r_{s,d} \leftarrow \underset{r \in \mathcal{R}}{\text{argmin}} \epsilon^{\text{out}}_{s,d}$; *//Select the best possible relay node $r_{s,d}$ and establish one cooperative link from the source (s) to the destination (d) to minimize the link outage $\epsilon^{\text{out}}_{s,d}$ according to* (2.11)

$R \leftarrow \{s, r_{s,d}, d\}$; *//Initial route established*

Updates:

if $\epsilon^{\text{out}}_{s,d} \leq \eta$; *//Compare with the target outage probability (constraint) η*

  $R \leftarrow \{s, r_{s,d}, d\}$; *//Route established*

else

  while $\vee \epsilon^{\text{out}}_{i,j} > \eta$, $\{i, r_{i,j}, j\} \in R$ **do** *//If any link $\epsilon^{\text{out}}_{i,j}$ along the constructed route is larger than the target error rate*

    $r_{i,r_{i,j}} \leftarrow \underset{r \in \mathcal{R}}{\text{argmin}} \epsilon^{\text{out}}_{i,r_{i,j}}$;

    $r_{r_{i,j},j} \leftarrow \underset{r \in \mathcal{R}}{\text{argmin}} \epsilon^{\text{out}}_{r_{i,j},j}$; *//New relay selections are triggered among the existing link in R to improve its $\epsilon^{\text{out}}_{i,j}$ performance*

    $R \leftarrow \{R - \{i, r_{i,j}, j\}\} \cup \{i, r_{i,r_{i,j}}, r_{i,j}\} \cup \{r_{i,j}, r_{r_{i,j},j}, j\}$; *//Update routing table*

  end

end if

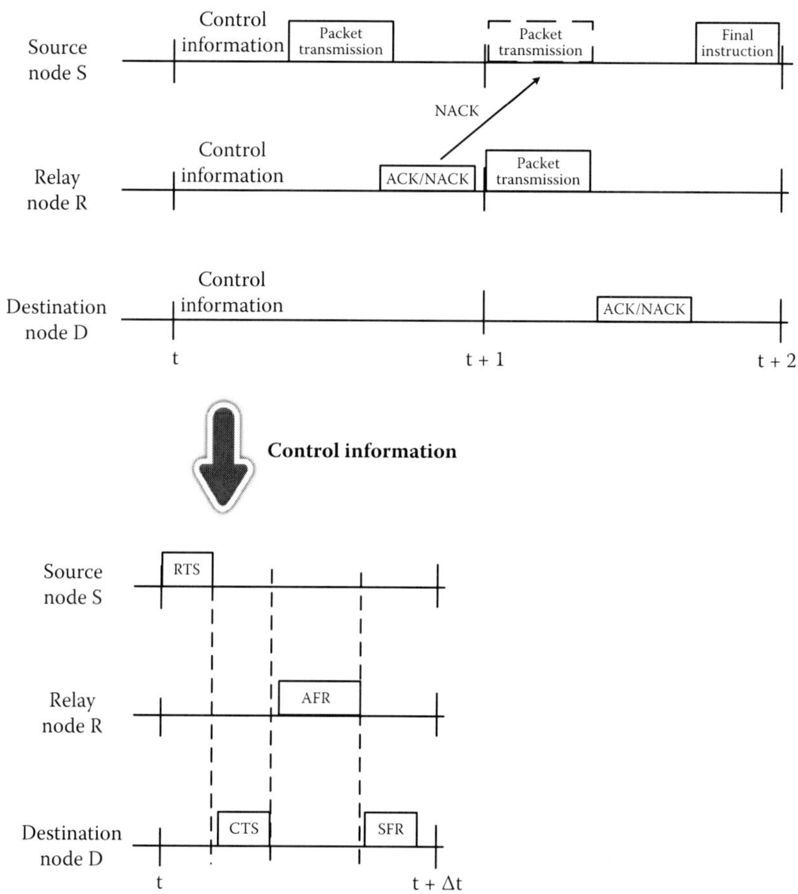

FIGURE 2.3 A timing diagram of cooperative transmission.

Theorem 2.1 *For an infinitely dense network where a node exists at any location, the end-to-end outage probability for the proposed routing with N hops is*

$$\epsilon_{\text{ETE}}^{\text{out}} \sim \Theta\left(\frac{1}{A^{2\alpha-1}}\right),$$

where A, being a perfect power of 2, is the largest integer that is smaller than the total number of hops N and α is the pass-loss exponent. The definition of $f(n) \sim \Theta(g(n))$ is that $\exists k_1, k_2 > 0, n_0, \forall n > n_0, |g(n)|k_1 \leq |f(n)| \leq |g(n)|k_2$.

Proof Suppose the total number of hops is N and the distance between the source and destination is D.

If $\log_2(N) = $ integer, then

$$\epsilon_{\text{ETE}}^{\text{out}} = N\frac{(2^{2b}-1)^2}{2p^2}\left(\frac{D}{N}\right)^{\alpha}\left(2\frac{D^{\alpha}}{2^{\alpha}N^{\alpha}}\right) = \frac{(2^{2b}-1)^2 D^{2\alpha}}{2^{\alpha}p^2N^{2\alpha-1}}. \tag{2.16}$$

Otherwise, determine the two nearest integers A and B that are next to N and satisfy $A < N < B$. Both A and B are perfect powers of 2. Therefore, we have

$$\epsilon_{\text{ETE}}^{\text{out}} = \frac{(2^{2b}-1)^2 D^{2\alpha}(N-A)}{2^{\alpha}p^2B^{2\alpha-1}A} + \frac{(2^{2b}-1)^2 D^{2\alpha}(2A-N)}{2^{\alpha}p^2A^{2\alpha-1}A}. \tag{2.17}$$

Using the relationship $B = 2A$, we have

$$\epsilon_{\text{ETE}}^{\text{out}} = \frac{(2^{2b}-1)^2 D^{2\alpha}}{2^{\alpha}\,p^2A^{2\alpha}}\left(\frac{N-A}{2^{2\alpha-1}} + 2A - N\right). \tag{2.18}$$

It is not difficult to observe that $\epsilon_{\text{ETE}}^{\text{out}} \sim \Theta\left(\frac{1}{A^{2\alpha-1}}\right)$. □

Motivated by such conclusion, we can find the performance of our proposed routing algorithm and optimal solution[1] in 2D infinitely dense networks, which are shown later in Figure 2.5. We observe that the proposed algorithm exhibits performance close to optimal, especially when the hop number N satisfies $\log_2(N) = $ integer.

[1]The optimal solution is defined as a route with no more N hops that minimizes the ETE outage performance in the cooperative networks. Detailed analysis is provided in Appendix A.1.

It is worth pointing out that we include an outage constraint in our proposed routing protocol for the following reasons: First, our proposed algorithm starts with routes with a small number of hops. Implicitly, it does not explore routes with an excessive number of hops. Instead, our algorithm achieves a good trade-off and balance between the hop count (which relates to delay) and ETE ϵ^{out} for routing, in order to achieve acceptable system performance. Second, for a given outage constraint, we can reduce the total number of nodes involved. Hence other benefits such as energy savings and communication traffic reducing could be realized.

Following the ideas above, we compare the minimum ETE ϵ^{out} achieved by our proposed routing algorithm with that of the optimal routing solution for a regularly dense linear network scenario. We consider a linear topology where nodes are located at equal distance from each other on a straight line. We assume that this distance between two adjunct nodes is D and the total number of nodes is n.

Before proceeding further, let us define a gap ratio g as the normalized difference between the outage probability for the best route established by our proposed algorithm and that of the optimal route

$$g = \frac{\epsilon_{\text{proposed}} - \epsilon_{\text{optimal}}}{\epsilon_{\text{optimal}}}. \tag{2.19}$$

The following theorem compares the performance of the routing algorithm to the optimal route.

Theorem 2.2 *For a regular linear network with n nodes ($\alpha = 2$),*

$$g = \begin{cases} 0, & \textit{if } \log_2(n-1) \textit{ or } \log_2(n) = \textit{integer}, \\ \dfrac{11}{4}, & \textit{if } \log_2\left(\dfrac{n-1}{3}\right) = \textit{integer}, \\ \dfrac{33}{2(n-1)}, & \textit{otherwise for an odd number of nodes.} \end{cases}$$

Proof See Appendix A.2. □

In general, Theorem 2.2 tells us that the proposed routing algorithm can have a ϵ^{out} close optimal. For example, for the first case where $n-1$ or n is a perfect power of 2, the proposed algorithm yields exactly the same ϵ^{out} as the optimal route. The gap ratio can be close to zero for the third case where

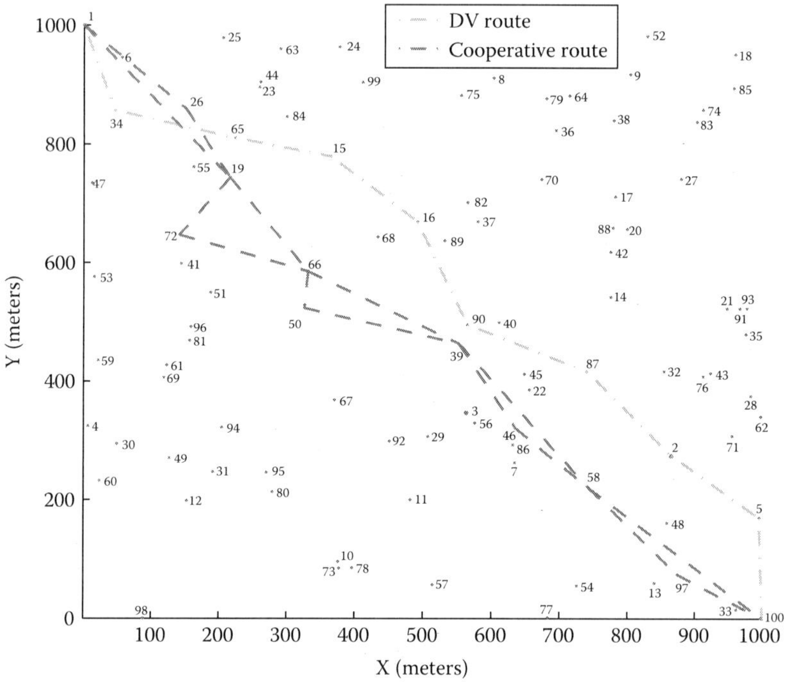

FIGURE 2.4 Routing comparison between proposed algorithm and destination-sequenced distance-vector (DSDV) algorithm.

the number of nodes is large enough. In addition to error performance, the proposed routing algorithm also provides the advantage of delay reducing. For example, for the second case, we can reduce $2^{\log_2 \frac{n-1}{3}} - 1$ hops and $\frac{n-1}{3}$ nodes involved when compared with optimal solution. For the third case, we can reduce one hop and two nodes involved.

2.5 SIMULATION RESULT

Figure 2.4 shows a routing example that is established by our proposed algorithm. The 100 nodes are uniformly distributed in 1000 m $\times$ 1000 m topology with the source and destination nodes located at the top left corner (node 1) and the bottom right corner (node 100), respectively. We set transmission power-to-noise ratio to 50 dB, data rate $b = 0.1$ bps/Hz, and the outage constraint $\epsilon^{\text{out}} = 0.01$. The light gray dashed line (located toward the upper right direction) is the distance-vector (DV) [21] routing, whereas the combined

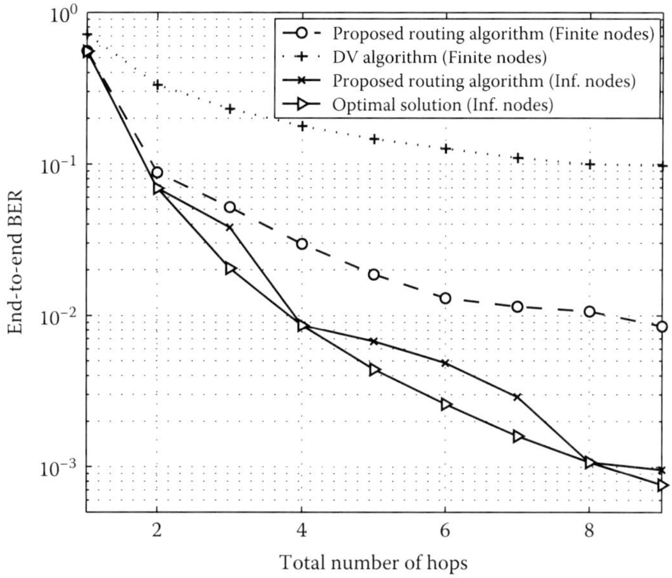

FIGURE 2.5 End-to-end BER versus total number of hops.

dark gray lines represent the proposed cooperative routing. For example, the cooperative link between node 1 and 19 uses node 26 as its relay. As shown in this figure, our proposed algorithm establishes a totally different route path compared with the DV routing algorithm. Furthermore, when compared with 9 hops and 10% end-to-end ϵ^{out} for the DV algorithm, the route generated by our proposed algorithm yields much better performance in terms of delay and outage probability: 5 hops and 3% end-to-end ϵ^{out}.

Moreover, under the same network assumption with a finite number of nodes, Figure 2.5 illustrates the ETE outage performance in terms of number of hops. It is shown that for cooperative routing, the ETE ϵ^{out} improves as the number of hops in the selected route increases. It also shows that cooperative transmission can achieve better ϵ^{out} performance than the DV algorithm. This implies that our proposed algorithm can generate routes with a smaller number of hops and satisfactory ETE ϵ^{out} when compared with the optimal solution from the DV algorithm. Such performance of the infinite node case can be treated as a low bound performance of the proposed algorithm.

CHAPTER 3

Energy Consumption of Cooperative Transmission

Zhengguo Sheng, University of British Columbia, Canada
Kin K. Leung, Imperial College London, UK

3.1 INTRODUCTION

The improvement on the ETE reliability is not the only benefit that we can get from cooperative routing. From another aspect, the total power consumption can also be reduced by using cooperation. Various cooperative routing algorithms have been developed to further reduce the total transmission power of cooperative transmission [22, 23, 24, 25, 26]. However, most existing cooperative routing algorithms are implemented by identifying a shortest path first, and thus the performance gains of cooperative communication cannot be fully exploited. Motivated by the QoS-driven routing algorithm in Section 2.4 and the optimal power allocation of DAF cooperative link in Section 7.1.1, we propose a power-efficient cooperative routing algorithm as follows.

The objective function of the power-efficient cooperative routing is similar to (2.15) and can be obtained as

$$p_{\text{ETE}} = \sum_{i,j \in S_1} p_D + \sum_{i,j \in S_2} (p^* + q^*), \tag{3.1}$$

where p_D is the power of direct transmission (2.6) and p^* and q^* are the optimal transmission power (7.6) of the source and relay, respectively.

3.2 DESCRIPTION OF THE POWER-EFFICIENT ROUTING ALGORITHM

Algorithm 3.1 describes the power-efficient routing algorithm, which is different from the routing algorithm in Section 2.4, in detail. Each node uses a default transmission power to construct a route at the initial stage. Since the optimal transmission power for both source and relay nodes can be determined by $d_{s,d}$, $d_{s,r}$, $d_{r,d}$, and link ϵ^{out}, each cooperative link can adjust to its minimum power in the mean time of the distributed relay selection once link outage performance has satisfied the target.

3.3 PERFORMANCE EVALUATION

In this section, we develop simulation results to illustrate the power savings of the power-efficient routing algorithm and then compare it with other cooperation-based power-saving algorithms.

We consider here a network scenario where a total number of N nodes are uniformly distributed in a 1000 m $\times$ 1000 m topology with the source and destination nodes located at the top left corner and the bottom right corner, respectively. Figure 3.1 shows the required total transmission power using different routing algorithms for different total numbers of nodes at the same link $\epsilon^{\text{out}} = 0.05$, $\alpha = 2$, and b = 0.95 bps/Hz. As shown, the total power consumption decreases as the network size increases. This is so because the distance between neighboring nodes is reduced with increased node density. Multi-hop routing ensures the lower power consumption between these nodes. We can also observe that the proposed power-efficient routing algorithm achieves the best performance among CASNCP [26], which is based on the shortest path algorithm, the conventional cooperative routing algorithm (with identical power assumption at both source and relay) 2.4, and the destination-sequenced distance-vector (DSDV) algorithm [21].

Since our proposed routing algorithm starts with routes with a small number of hops, Figure 3.2 shows the relationship between the total power consumption of the cooperative route in terms of total number of hops and its ETE outage performance.

ALGORITHM 3.1 Power-Efficient Routing Algorithm

Input: Relay candidates $\mathcal{R}$, default transmission power p, data rate b, outage constraint η

Output: Cooperative route R, link outage ϵ^{out}, and optimal transmission power (p^* and q^*)

Initialize:

$r_{s,d} \leftarrow \underset{r \in \mathcal{R}}{\text{argmin}} \epsilon^{\text{out}}_{s,d}$; *//Select the best possible relay node $r_{s,d}$ and establish one cooperative link from the source (s) to the destination (d) to minimize the link outage $\epsilon^{\text{out}}_{s,d}$ according to* (2.11)

$R \leftarrow \{s, r_{s,d}, d\}$; *//Initial route established*

Updates:

if $\epsilon^{\text{out}}_{s,d} \leq \eta$; *//Compare with the target outage probability (constraint) η*

  $R \leftarrow \{s, r_{s,d}, d\}$; *//Route established*

  $p_s \leftarrow p^*$, $q_{r_{s,d}} \leftarrow q^*$; *//Update transmission power*

else

  while $\vee \epsilon^{\text{out}}_{i,j} > \eta$, $\{i, r_{i,j}, j\} \in R$ **do** *//If any link $\epsilon^{\text{out}}_{i,j}$ along the constructed route is larger than the target error rate*

    $r_{i,r_{i,j}} \leftarrow \underset{r \in \mathcal{R}}{\text{argmin}} \epsilon^{\text{out}}_{i,r_{i,j}}$;

    $r_{r_{i,j},j} \leftarrow \underset{r \in \mathcal{R}}{\text{argmin}} \epsilon^{\text{out}}_{r_{i,j},j}$; *//New relay selections are triggered among the existing link in R to improve its $\epsilon^{\text{out}}_{i,j}$ performance*

    $R \leftarrow \{R - \{i, r_{i,j}, j\}\} \cup \{i, r_{i,r_{i,j}}, r_{i,j}\} \cup \{r_{i,j}, r_{r_{i,j},j}, j\}$; *//Update routing table*

  end

  for all $\vee \{i, r_{i,j}, j\} \in R$ **do**

  $p_i \leftarrow p^*$, $q_{r_{i,j}} \leftarrow q^*$; *//Update transmission power*

  $p_{r_{i,j}} \leftarrow p^*$, $q_j \leftarrow q^*$; *//Update transmission power*

  end for

end if

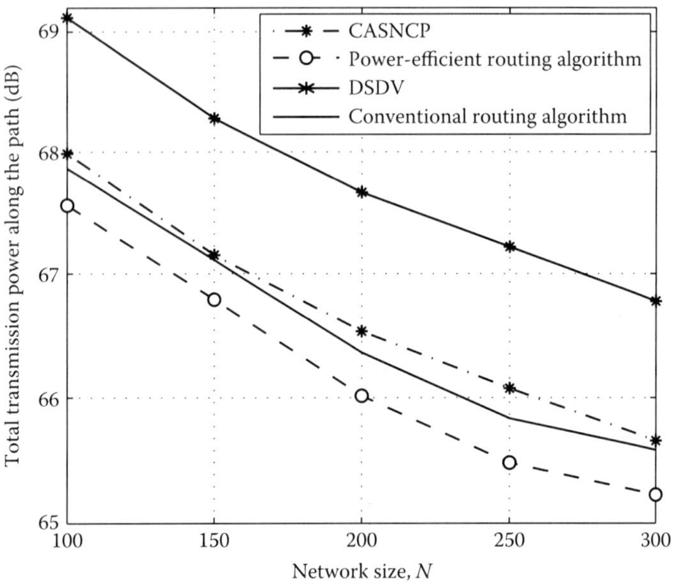

FIGURE 3.1 Network size versus total normalized transmission power consumption along the path.

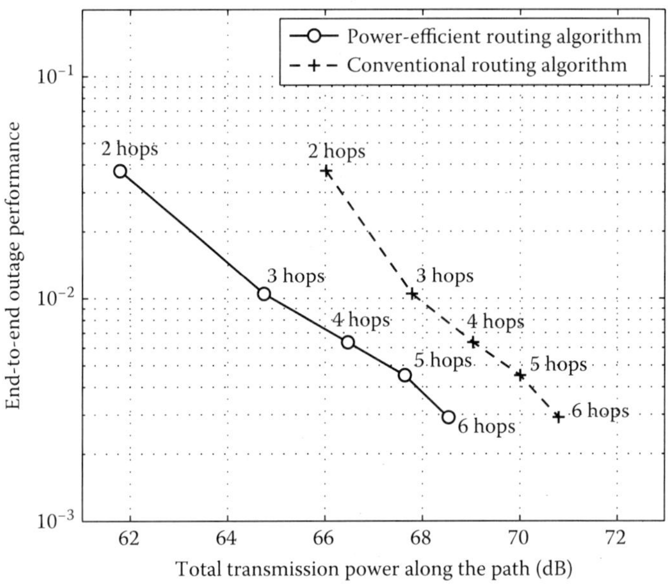

FIGURE 3.2 Total normalized transmission power along the path versus end-to-end outage performance.

The network scenario is the same as in Figure 3.1, but with a fixed $N = 100$, $\alpha = 2$, and $b = 0.2$ bps/Hz. The total transmission power along the path is proportional to total number of hops. Under the same network topology, as the total number of hops increases, the total transmission power is increased. Meanwhile, under the same route and ETE outage achievement, the proposed algorithm can reduce the total power consumption by a couple dB.

CHAPTER 4

Throughput of Cooperative Transmission

Zhengguo Sheng, University of British Columbia, Canada
Zhiguo Ding, Newcastle University, UK

4.1 INTRODUCTION

In wireless networks, the broadcast nature of wireless transmission enables cooperation by sharing the same transmissions with nearby receivers and thus can help improve spatial reuse and boost network throughput along a multi-hop routing. The performance of wireless networks can be further improved if prior information available at the receivers can be utilized to achieve perfect interference subtraction. In this section, we investigate performance gain on network throughput for wireless cooperative networks by using a simple multiuser detection (MUD) scheme, called overlapped transmission, in which multiple transmissions are allowed only when the information in the interfering signal is known at the receiver.

The idea of employing MUD in wireless networks to increase spatial reuse and throughput has been proposed [27, 28, 29, 30]. We have learned from existing works that typical MUD schemes (e.g., Successive Interference Cancellation (SIC)) need significant process power. However, for wireless ad hoc networks, it may not be possible. Motivated by the fact that prior infor-

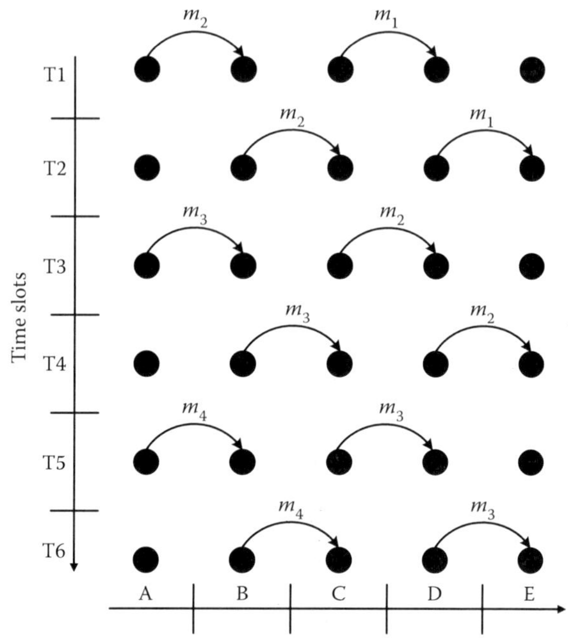

FIGURE 4.1 Multi-hop direct transmission with overlapping in a five-node linear network.

mation available at the receiver can be utilized to achieve perfect interference subtraction by using the MUD scheme [30] and therefore invite more simultaneous transmissions along a multi-hop routing, we propose here to further exploit network throughput in cooperative networks by combining the MUD scheme with supplementary cooperation strategy.

4.2 INTERFERENCE SUBTRACTION IN A MULTI-HOP SCENARIO

In this section, we illustrate the idea of interference subtraction in a five-node linear network shown in Figures 4.1 and 4.2. Without loss of generality as in the previous work [30], the distance between adjacent nodes is 1, the transmission range (solid line) is also assumed to be 1, and the interference range (dashed line) is assumed to be twice the transmission range. Specifically, we use outage probability to define the transmission range and interference range.

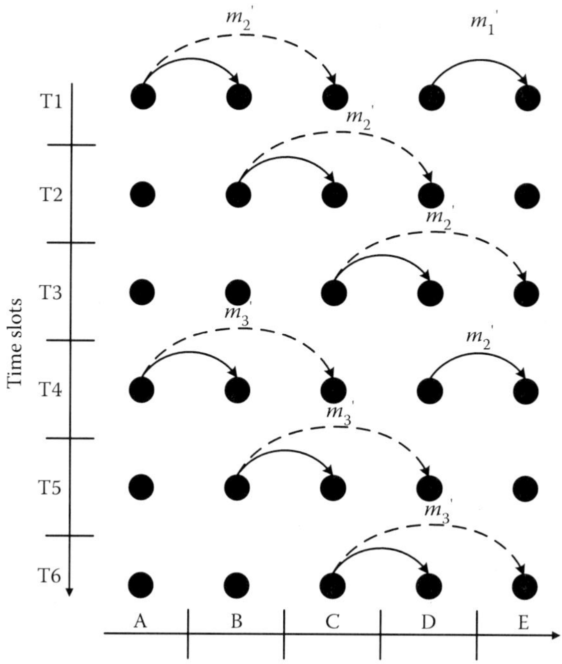

FIGURE 4.2 Multi-hop cooperative transmission with overlapping in a five-node linear network.

We still employ the same propagation model in Chapter 2 to consider path-loss and Rayleigh fading. The wireless link $a_{i,j}$ between the nodes i and j is modeled as $a_{i,j} = h_{i,j}/d_{i,j}^{\alpha/2}$, where $d_{i,j}$, the distance between the nodes i and j, represents the large-scale behavior of the channel gain, α is the path-loss exponent, and $h_{i,j}$ is assumed to be the independent and identically distributed (i.i.d.) complex Gaussian variable with zero mean and unit variance.

For direct transmission, according to (2.5), the outage probability satisfies

$$\epsilon_D^{\text{out}} = \Pr[I_D < b] = d_{s,d}^{\alpha}\left(\frac{2^b - 1}{p}\right), \tag{4.1}$$

where b is the desired data rate in bps/Hz, which is defined by the QoS requirement, and d is the distance between two nodes.

By using (4.1), we have the equivalent definition as follows:

$$\begin{cases} \text{node is within transmission range,} & \text{if} \quad \epsilon_D^{\text{out}} \leq \dfrac{2^b - 1}{p}, \\ \text{node is within interference range,} & \text{if} \quad \epsilon_D^{\text{out}} \leq \dfrac{2^\alpha(2^b - 1)}{p}, \\ \text{interference free,} & \text{if} \quad \epsilon_D^{\text{out}} > \dfrac{2^\alpha(2^b - 1)}{p}. \end{cases} \tag{4.2}$$

Therefore, when setting a desired data rate b and carefully choosing a transmission power p, a successful transmission can be made only if the outage probability at the receiver satisfies the first condition of (4.2). When a node is within the interference range, which satisfies the second condition, it cannot directly decode the message from the source. However, from information theory's perspective, it can accumulate the information from both the source and relay to satisfy the first condition by using cooperative transmission in two time slots, which is shown in Figure 4.2.

For cooperative transmission, let $d_{s,d}$, $d_{s,r}$, and $d_{r,d}$ be the respective distances among the source, relay, and destination of one single cooperative link. The outage probability is

$$\epsilon_C^{\text{out}} = \frac{1}{2} d_{s,d}^\alpha (d_{s,r}^\alpha + d_{r,d}^\alpha) \frac{(2^{b_C} - 1)^2}{p^2}. \tag{4.3}$$

Note that the mathematical details behind this equation are omitted and can be found in Section 2.2.2.

Network throughput can be improved by employing simultaneous transmission [30], and the scheduling scheme employing the overlapped transmission for the five-node linear network is depicted in Figure 4.1. We observe that in time slot T3, node C forwards packet m_2, which is received by node B in T2, to node D. Node B can actually keep a copy of the transmitted message m_2 locally; thus it knows the message being transmitted by node C in T3 and can apply the MUD with the stored prior information m_2 to mitigate the interference caused by node C, while node A is allowed to transmit another message m_3 at the same time.

The performance of the scheduling schemes is measured in terms of network throughput at destination E. We assume time slots are of equal length T and identical transmission power for all nodes. Since destination E successfully receives a message on average in every two time slots, the average throughput for direct transmission with overlapping is

$$\lambda_D = \frac{b}{2}. \tag{4.4}$$

Under the same outage achievement $\epsilon_C^{\text{out}} = 1-(1-\epsilon_D^{\text{out}})^2$, by using (4.1) and (4.3), the data rate for cooperative transmission can be increased to

$$b_C \approx b + \log_2\left(\sqrt{\frac{2}{\epsilon_{TH}}}\right), \tag{4.5}$$

where ϵ_{TH} is the outage probability of a two-hop-length direct transmission. For cooperative transmission as shown in Figure 4.2, a message on average requires three time slots to be received at destination E; the average throughput for cooperative transmission with overlapping is

$$\lambda_C = \frac{b_C}{3} = \frac{b + \log_2\left(\sqrt{\frac{2}{\epsilon_{TH}}}\right)}{3}. \tag{4.6}$$

As a result, the performance of cooperative transmission with overlapping is better than that of direct transmission with overlapping only when $\lambda_D < \lambda_C$, which equals

$$\epsilon_{TH} < \frac{2}{(2^{b/2})^2}. \tag{4.7}$$

We observe that (4.7) can be easily satisfied, especially when the transmission power-to-noise ratio p is large enough. Let us consider an example. We assume $\alpha = 2$, $b = 2$ bps/Hz, and $p = 20$ dB; then with the same outage performance, $b_C = 3.73$ bps/Hz. The transmission efficiency, which is defined as the ratio of network throughput of cooperative transmission scheduling employing overlapped transmission to that of direct transmission scheduling employing overlapped transmission, is $\Gamma = \frac{\lambda_C}{\lambda_D} = 1.24$.

It is clear that the scheduling scheme of cooperative transmission with overlapped transmission has shown better potential to improve network throughput by 24% over the scheme of direct transmission with overlapped transmission and that potential can be further improved when one implements supplementary cooperation with overlapped transmission, as will be examined in the next section.

4.3 SUPPLEMENTARY COOPERATION

In this section, we introduce another idea of supplementary cooperation strategy. We have focused so far on the conventional cooperation strategy that the mutual information accumulation only happens at the destination node of each cooperative link. Actually, the relay node can also get full benefits from cooperation by taking advantage of the broadcast nature of wireless transmission to further reduce the decoding error. As depicted in Figure 4.2, in the second time slot, node C receives the second copy of m_2' from relay B. At the same time, relay D can actually overhear the same m_2' (dashed line). That is so because the node D is within the interference range of node B and the same packet needs to go through all the nodes along the route. In essence, nodes B, C, and D can consist of another cooperative link called *supplementary cooperation*.

Consider a general network scenario where simultaneous transmissions are among the same route, which is shown in Figure 4.3. As an extension, we are interested in the interference impact on network performance, that is, under a realistic assumption that multiple nodes are active for transmission at the same time. Since we assume that each node uses the same transmission power, the single-to-interference-plus-noise ratio (SINR) at receiver s_{i+1} is

$$p' = \text{SINR} = \frac{p}{p_0 + p_I} = \frac{p}{p_0 + \sum_j p|\text{a}_{j,s_{i+1}}|^2}, \tag{4.8}$$

where p_I is the summation of interfering power at the receiver. For Rayleigh fading, $|\text{a}_{j,s_{i+1}}|^2$ is exponentially distributed with parameter $d^{\alpha}_{j,s_{i+1}}$. By taking the average of $|\text{a}_{j,s_{i+1}}|^2$ and assuming the white noise power $p_0 \ll p_I$, then the above SINR is

$$p' = \text{SINR} \approx \frac{p}{p_I} = \frac{p}{\sum_j p/d^{\alpha}_{j,s_{i+1}}} = \frac{1}{\sum_j 1/d^{\alpha}_{j,s_{i+1}}}. \tag{4.9}$$

Hence the mutual information at node s_{i+1} can be shown as follows:

$$I = \begin{cases} \frac{1}{2}\log(1 + 2p'_{s_{i+1}}|\text{a}_{s_i,s_{i+1}}|^2), & \text{if } |\text{a}_{r_{i-1},r_i}|^2 + |\text{a}_{s_i,r_i}|^2 < G(p'_{r_i}), \\ \frac{1}{2}\log(1 + p'_{s_{i+1}}|\text{a}_{s_i,s_{i+1}}|^2 & \text{if } |\text{a}_{r_{i-1},r_i}|^2 + |\text{a}_{s_i,r_i}|^2 \geq G(p'_{r_i}), \\ \quad + p'_{s_{i+1}}|\text{a}_{r_i,s_{i+1}}|^2), & \end{cases} \tag{4.10}$$

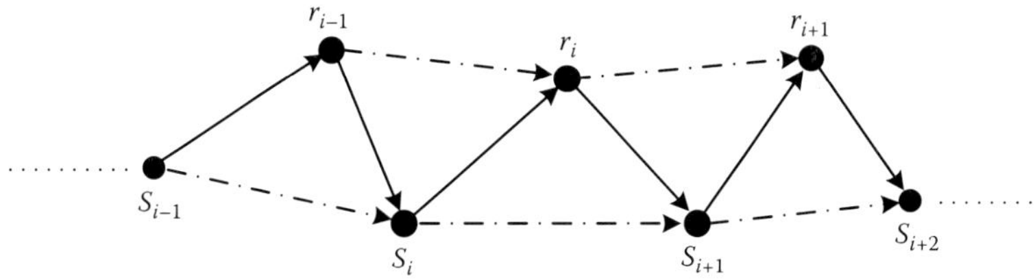

FIGURE 4.3 One example of a cooperative route.

where p'_{r_i} and $p'_{s_{i+1}}$ are transmission power-to-interference ratios at relay r_i and destination s_{i+1}, respectively, and $G(p'_{r_i}) = (2^{2b} - 1)/p'_{r_i}$. The first case in (4.10) corresponds to relay r_i that is not able to decode through supplementary cooperation with r_{i-1} and s_i, and thus source s_i is repeating its transmission. The maximum average mutual information is that of repetition coding from source s_i to destination s_{i+1}; therefore, the extra factor of 2 is added in the SINR. The second case corresponds to relay r_i that has the ability to decode and repeat the transmission through supplementary cooperation; then the maximum average mutual information is that repetition coding from both s_i and r_i to destination s_{i+1}.

Therefore, the outage event for such DAF is given by $I < b$ and is equivalent to the event

$$\left(\{|\mathrm{a}_{r_{i-1},r_i}|^2 + |\mathrm{a}_{s_i,r_i}|^2 < G(p'_{r_i})\} \cap \{2|a_{s_i,s_{i+1}}|^2 < G(p'_{s_{i+1}})\}\right) \cup$$

$$\left(\{|\mathrm{a}_{r_{i-1},r_i}|^2 + |\mathrm{a}_{s_i,r_i}|^2 \geq G(p'_{r_i})\} \cap \{|a_{s_i,s_{i+1}}|^2 + |a_{r_i,s_{i+1}}|^2 < G(p'_{s_{i+1}})\}\right). \tag{4.11}$$

As can be seen, two events of the union in (4.11) correspond to two cases in (4.10), respectively. Because the events in union of (4.11) are mutually exclusive, the outage performance of s_{i+1} with supplementary cooperation (SC) becomes

$$\begin{aligned}
\epsilon_{SC}^{\text{out}} &= \Pr[I < b] \\
&= \underbrace{\Pr[|\mathrm{a}_{r_{i-1},r_i}|^2 + |\mathrm{a}_{s_i,r_i}|^2 < G(p'_{r_i})]}_{\text{T1}} \underbrace{\Pr[2|\mathrm{a}_{s_i,s_{i+1}}|^2 < G(p'_{s_{i+1}})]}_{\text{T2}} \\
&\quad + \underbrace{\Pr[|\mathrm{a}_{r_{i-1},r_i}|^2 + |\mathrm{a}_{s_i,r_i}|^2 \geq G(p'_{r_i})]}_{\text{T3}} \\
&\quad \times \underbrace{\Pr[|\mathrm{a}_{s_i,s_{i+1}}|^2 + |\mathrm{a}_{r_i,s_{i+1}}|^2 < G(p'_{s_{i+1}})]}_{\text{T4}}.
\end{aligned} \tag{4.12}$$

Here, we compute a closed form for (4.12). By computing the large SNR behavior, we have the limits

$$\text{T1} \longrightarrow \frac{1}{2} d^{\alpha}_{r_{i-1},r_i} d^{\alpha}_{s_i,r_i} G^2(p'_{r_i}), \qquad \text{T2} \longrightarrow \frac{1}{2} d^{\alpha}_{s_i,s_{i+1}} G(p'_{s_{i+1}}),$$
$$\text{T3} \longrightarrow 1 - \frac{1}{2} d^{\alpha}_{r_{i-1},r_i} d^{\alpha}_{s_i,r_i} G^2(p'_{r_i}), \quad \text{T4} \longrightarrow \frac{1}{2} d^{\alpha}_{s_i,s_{i+1}} d^{\alpha}_{r_i,s_{i+1}} G^2(p'_{s_{i+1}}).$$

Then, (4.12) equals

$$\begin{aligned} P^{\text{out}}_{SC} &= \Pr[I < b] \\ &= \frac{1}{4} d^{\alpha}_{r_{i-1},r_i} d^{\alpha}_{s_i,r_i} d^{\alpha}_{s_i,s_{i+1}} G^2(p'_{r_i}) G(p'_{s_{i+1}}) \\ &\quad + \frac{1}{2} d^{\alpha}_{s_i,s_{i+1}} d^{\alpha}_{r_i,s_{i+1}} G^2(p'_{s_{i+1}}) \left(1 - \frac{1}{2} d^{\alpha}_{r_{i-1},r_i} d^{\alpha}_{s_i,r_i} G^2(p'_{r_i})\right), \end{aligned} \tag{4.13}$$

where

$$G(p'_{s_{i+1}}) = (2^{2b} - 1) \sum_j \frac{1}{d^{\alpha}_{j,s_{i+1}}} \quad \text{and} \quad G(p'_{r_i}) = (2^{2b} - 1) \sum_j \frac{1}{d^{\alpha}_{j,r_i}}.$$

Under the same network scenario as depicted in Figure 4.2, (4.13) can be simplified as

$$\epsilon^{\text{out}}_{SC} = \frac{d^{\alpha}_{s,d} d^{\alpha}_{r,d} (2^{b_{SC}-1})^2}{2p^2} + \frac{d^{\alpha}_{s,d} d^{\alpha}_{s,r} d^{\alpha}_{r',r} (2^{b_{SC}-1})^3}{4p^3}, \tag{4.14}$$

where $d_{r',r}$ is the distance between two adjacent relay nodes. Then we have the data rate of SC

$$b_{SC} = \log_2(xp + 1), \tag{4.15}$$

where

$$x = -\frac{B}{3u} + u - \frac{A}{3}, \quad A = \frac{1}{2^{\alpha-1}}, \quad B = -\frac{1}{3 \cdot 2^{2\alpha-2}},$$
$$u = \sqrt[3]{-\frac{q}{2} - \sqrt{\frac{q^2}{4} + \frac{B^3}{27}}}, \quad \text{and} \quad q = -\frac{\epsilon^{\text{out}}_{SC}}{2^{2\alpha-2}} + \frac{2}{27 \cdot 2^{3\alpha-3}}.$$

Consider the same example in Section 4.2; we get $b_{SC} = 4.25$ bps/Hz and the average throughput for supplementary cooperation is $\lambda_{SC} = \frac{b_{SC}}{3}$. Then the transmission efficiency is $\Gamma' = \frac{\lambda_{SC}}{\lambda_D} = 1.42$.

In general, the results from above tell us that the supplementary cooperation with overlapped transmission achieves the best performance among the three schemes. It is worth noting that the supplementary cooperation can be realized simply by taking advantage of the broadcast nature of wireless transmission. Hence, compared with conventional cooperative transmission in Section 2.2.2, there is no extra system overhead involved.

4.4 SIMULATION RESULT

In this section, we evaluate the performance of supplementary cooperation. Especially, we employ the space-time reuse scheme to analyze the interference impact on network throughput for later comparison.

We consider here a network scenario that 100 nodes are uniformly distributed in a 1000 m $\times$ 1000 m topology with the source and destination nodes located at the top left corner and the bottom right corner, respectively. We set the transmission power-to-noise ratio as 60 dB and desired data rate $b = 0.2$ bps/Hz. Results are averaged over 100 simulation runs. By using the cooperative routing algorithm in Section 2.4, Figure 4.4 reports ETE outage performance of routes with different total number of hops from the source to destination. In this simulation, we consider the simplest case where only one transmission is possible in each time slot. So there is no other interference present. As can be seen, the supplementary cooperation achieves much better error performance than the conventional cooperation as well as the same route with only direct transmission. Especially, an average outage reduction of 34.87% is achieved when compared with the conventional cooperation. In addition, it further shows that for cooperative routing, ETE outage improves as the number of hops in the selected route increases. In particular, we observe that 3 hops supplementary cooperation already has the better ETE outage performance than 4 hops conventional cooperation. This implies that supplementary cooperation can generate routes with a smaller number of hops and satisfactory ETE outage when compared with the conventional cooperation.

Next, we evaluate ETE outage performance of supplementary cooperation with overlapped transmission under an interfering environment, which

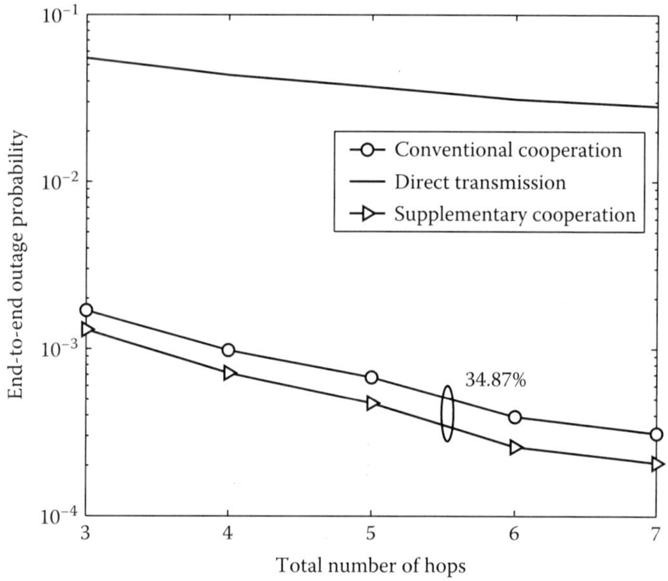

FIGURE 4.4 End-to-end outage performance versus total number of hops.

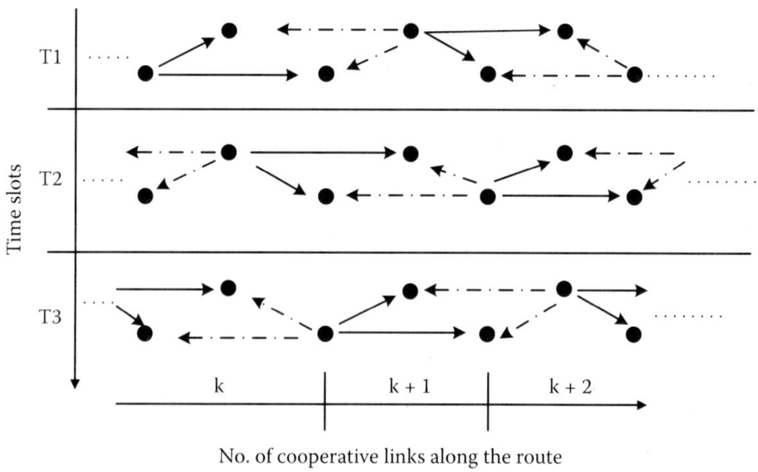

FIGURE 4.5 TDM-schedule for a cooperative route with $M = 1$.

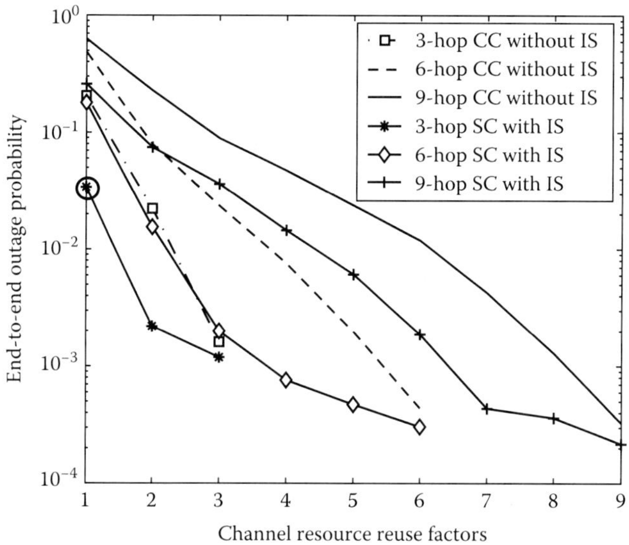

FIGURE 4.6 End-to-end outage performance versus channel resource reuse factors.

allows multi-node transmissions along the same routing using the space-time reuse scheme. We assume all nodes along a route transmit in the same frequency band and employ a regular time-division multiplex schedule (TDM-schedule) of length M-cooperative-links so that in time slot t, the nodes $2iM + (t \bmod 2M)$ are allowed to transmit, for $i = ... - 1, 0, 1....$ Because each cooperative link consists of two transmissions from the source and relay in two consecutive time slots, the extra factor 2 is added in TDM-schedule. Figure 4.5 shows the TDM-schedule for a general cooperative route with reuse factor $M = 1$. The solid lines are simultaneous transmissions and the dashed lines are interferences that can be canceled through overlapped transmissions.

Figure 4.6 shows ETE outage performance of supplementary cooperation (SC) with interference subtraction (IS) by using the overlapped transmission and conventional cooperation (CC) without IS. It is clear that supplementary cooperation with interference subtraction achieves much better performance than the conventional way. Furthermore, a careful reader might notice that in fact there is a trade-off between reuse factor, ETE outage probability, and network throughput. As reuse factor increases, ETE outage probability is reduced correspondingly. However, the network throughput is adversely affected by large reuse factor. Therefore, in order to find the best cooperative route achieving maximum network throughput, we define the network throughput as follows:

$$\lambda = \begin{cases} \dfrac{b(1-\epsilon_{\mathrm{ETE}})}{2M+1}, & \text{if } M = 1, \\ \dfrac{b(1-\epsilon_{\mathrm{ETE}})}{2M}, & \text{if } M > 1, \end{cases} \tag{4.16}$$

where b is the desired data rate, ϵ_{ETE} is ETE outage probability, and M is the reuse factor. By using (4.16), we find that 3 hops supplementary cooperation with interference subtraction is the best routing to achieve the maximum throughput in such a network scenario, which is circled in Figure 4.6.

CHAPTER 5

Delay Analysis of Cooperative Transmission

Zhiguo Ding, Newcastle University, UK
Kin K. Leung, Imperial College London, UK
Zhengguo Sheng, University of British Columbia, Canada

5.1 INTRODUCTION

A large ETE throughput does not necessarily result in a small ETE transmission delay. The size of each packet and type of transmission schedule also play important roles in network performance. In this section, we employ an error exponent model [31] to study the transmission time of wireless networks using decode-and-forward (DAF), amplify-and-forward (AF), and multi-hop (MH) cooperative protocols.

5.2 SYSTEM MODEL AND DELAY BEHAVIORS

Each node in the network is equipped with one omnidirectional antenna element. The Time Division Multiple Access (TDMA) scheme is used to enable various nodes to share the same frequency band. We employ a channel model incorporating path-loss and additive white Gaussian noise [32]. The received signal at node j is modeled as

$$y_j = \mathrm{a}_{i,j} x_i + n_j, \tag{5.1}$$

where x_i is the signal transmitted by node i and n_j is additive white Gaussian noise, with variance σ_n^2, at the receiver. The channel gain $\mathrm{a}_{i,j}$ between the nodes i and j is modeled as $\mathrm{a}_{i,j} = 1/d_{i,j}^{\alpha/2}$, where $d_{i,j}$ is the distance between the nodes i and j and α is the path-loss exponent.

For existing encoders, it is inevitable to output some bits that are correlated to each codeword and therefore cause the loss of independence. Since random Gaussian inputs can maximize the mutual information (or entropy), the use of random Gaussian code can theoretically ensure the achievability of the optimal performance and therefore maximize cooperative diversity. We assume that each node uses a random Gaussian code[1] to encode a block of L nats[2] information into a time signal of infinite duration and transmit it. However, the transmitter will transmit a finite time only until the receiver successfully decodes the message; thus only a finite length of codeword will be transmitted. Since the Gaussian waveform channel can be modeled as a sequence of complex Gaussian scalar channels [33, 31, 34], if we use the output of the first N channels to decode the transmitted message (i.e., decoding at time N/W, where N is number of samples used for decoding and W is bandwidth), the coding bound on block error probability ϵ is

$$\epsilon \leq \exp\left(\rho L - \sum_{i=1}^{N} (E_0(\rho, \mathrm{SINR}_i))\right). \tag{5.2}$$

Assuming that each node transmits in the same frequency band (with normalized bandwidth $W = 1$) and the signal-to-interference-plus-noise ratio SINR at the receiver keeps the same during the one block transmission, we can simplify (5.2) as

$$\epsilon \leq \exp(\rho L - N(E_0(\rho, \mathrm{SINR})) \tag{5.3}$$

[1] The Gaussian code encoder separates the incoming binary data stream into equal lengths of L binary digits each. There are in total $M = 2^L$ different binary sequences of length L, and the encoder provides a codeword for each. Each codeword is a sequence of a fixed number, N, of channel input letters. The codewords are samples of bandlimited white Gaussian noise.

[2] In order to simplify notation and analysis, we use information unit *nat* in this paper; 1 nat $= \log_2 e$ bit.

for any constant factor $\rho \in [0, 1]$. $E_0(\rho, \text{SINR})$ is the error exponent determined by ρ and SINR. For a complex Gaussian channel with unit bandwidth, a simple expression for the error exponent [31] is derived from

$$E_0(\rho, \text{SINR}) = \rho \ln \left(1 + \frac{\text{SINR}}{1+\rho}\right). \tag{5.4}$$

Given a target block error probability ϵ that the receiver can successfully decode the message, the minimum coding length N is bounded by

$$N \geq \frac{\rho L - \ln \epsilon}{\rho \ln \left(1 + \frac{\text{SINR}}{1+\rho}\right)}. \tag{5.5}$$

The lower bound is the minimum coding length for sending L nats information over one transmission link when a target reliability constraint is guaranteed. Given that the decoding time is $D = N/W$, we will use this lower bound as the minimum delay to characterize delay performance of three low-complexity cooperative protocols that can be utilized in the network of Figure 2.1, including *amplify-and-forward (AF)*, *decode-and-forward (DAF)*, and *multi-hop (MH)* [35]. As a result, the destination using AF or DAF receives two independent copies of the same packets transmitted through different wireless channels, from which diversity gain can be achieved, whereas the destination using MH only receives one copy from its relay node.

5.2.1 Amplify-and-Forward Transmission

The relay node amplifies whatever it has received subject to its power constraint and retransmits the signals to the destination in the second time slot. As explained in detail in Laneman et al. [1], the average channel capacity between the source and the destination is given by

$$I_{\text{AF}} = \frac{1}{2} \log(1 + \text{SINR}_{s,d} + f(\text{SINR}_{s,r}, \text{SINR}_{r,d})), \tag{5.6}$$

where SINR is defined as received power-to-noise-plus-interference ratio and $f(x, y) = \frac{xy}{x+y+1}$. From (5.5), the delay performance of AF can be obtained as

$$D_{\mathrm{AF}} \geq \frac{2(\rho L - \ln \epsilon)}{\rho \ln\left(1 + \frac{\mathrm{SINR}_{s,d} + f(\mathrm{SINR}_{s,r}, \mathrm{SINR}_{r,d})}{1+\rho}\right)}. \tag{5.7}$$

Note that the source and relay transmit an identical codeword in two equal time slots; the extra factor of 2 is added in the delay.

5.2.2 Decode-and-Forward Transmission

Let $d_{s,d}$, $d_{s,r}$, and $d_{r,d}$ be the respective distances among the source, relay, and destination. During the first time slot, the destination receives $y_d = (1/d_{s,d}^{\alpha/2})x_s + n_d$ from the source node, where x_s is the information transmitted by the source and n_d is white noise. During the second time slot, the destination node receives

$$y_d = \begin{cases} \frac{1}{d_{s,d}^{\alpha/2}} x_s + n_d, & \text{if } b > I_{s,r}, \\ \frac{1}{d_{r,d}^{\alpha/2}} x_r + n_d, & \text{if } b \leq I_{s,r}, \end{cases} \tag{5.8}$$

where $b = \frac{L}{2N \ln 2}$ bps/unit hertz,[3] N is the coding length, and $I_{s,r} = \frac{1}{2}\log(1 + \mathrm{SINR}_{s,r})$ can be derived from direct transmission. In this protocol, the relay transmits only if the desired data rate b is below the channel capacity; otherwise, the source retransmits in the second time slot. We thus implicitly assume a mini-slot at the beginning of the second slot during which ACKs are sent error-free from relay to source.

Assuming that the relay node can perform perfect decoding, the channel capacity of this cooperative link can be shown to be

$$I_{\mathrm{DAF}} = \begin{cases} \frac{1}{2}\log(1 + 2\mathrm{SINR}_{s,d}), & \text{if } b > I_{s,r}, \\ \frac{1}{2}\log(1 + \mathrm{SINR}_{s,d} + \mathrm{SINR}_{r,d}), & \text{if } b \leq I_{s,r}. \end{cases} \tag{5.9}$$

Note that the same noise variance is assumed at both relay and destination.

---

[3] Since we use the information nat as the unit in this paper, the original data rate $b = \frac{L}{NW\tau_t}$ in nat/s/unit hertz should be converted to that in bit/s/unit hertz. The sampling time τ_t at the decoder equals Nyquist rate W of 1 unit time per symbol.

Therefore, the delay performance of DAF transmission is shown as

$$D_{\text{DAF}} \geq \begin{cases} 2\dfrac{\rho L - \ln \epsilon}{\rho \ln\left(1 + \frac{2\text{SINR}_{s,d}}{1+\rho}\right)}, & \text{if } b > I_{s,r}\,, \\ 2\dfrac{\rho L - \ln \epsilon}{\rho \ln\left(1 + \frac{\text{SINR}_{s,d} + \text{SINR}_{r,d}}{1+\rho}\right)}, & \text{if } b \leq I_{s,r}\,, \end{cases} \tag{5.10}$$

where $b = \frac{L}{2N \ln 2}$. It is worth noting that the data rate b is also determined by the coding length N, so one simple way to determine the minimum delay of DAF is to calculate the two delays (or coding length) using (5.10) and then bring them back to validate the conditions.

5.2.3 Multi-hop Transmission

Different from AF and DAF, multi-hop has the source transmitting its signals to the relay in one time slot, and then the relay forwarding the signals to the destination in a second time slot. In order to derive its delay performance, we formulate an optimization problem to minimize the link delay with a constrained block error probability ϵ as is shown:

$$\begin{aligned} \min \quad & D_{s,r} + D_{r,d} \\ \text{s.t.} \quad & 1 - (1 - \epsilon_{s,r})(1 - \epsilon_{r,d}) \leq \epsilon\,. \end{aligned} \tag{5.11}$$

Then the delay performance of two-hop can be derived as

$$D_{\text{MH}} \geq \frac{\rho L - \ln\left(\frac{\epsilon K_{s,r}}{K_{s,r} + K_{r,d}}\right)}{\rho \ln\left(1 + \frac{\text{SINR}_{s,r}}{1+\rho}\right)} + \frac{\rho L - \ln\left(\frac{\epsilon K_{r,d}}{K_{s,r} + K_{r,d}}\right)}{\rho \ln\left(1 + \frac{\text{SINR}_{r,d}}{1+\rho}\right)}, \tag{5.12}$$

where $K_{i,j} = 1/\rho \ln(1 + \frac{\text{SINR}_{i,j}}{1+\rho})$. The development of (5.12) is similar to that in Section 5.3; refer to Appendix A.3.

To illustrate the delay performance, we provide an example to show the minimum delay achieved by different cooperative protocols. Consider a 10 m $\times$ 10 m network with a center at (0, 0) and a source and destination located at (5 m, 0) and (-5 m, 0), respectively. One hundred relay candidates are uniformly distributed within the network. The transmission power-to-noise ratio is assumed to be 10 dB, path-loss exponent is set as $\alpha = 3$, $\rho = 0.5$, and the prefixed error probability is $\epsilon = 0.001$. As can be seen from

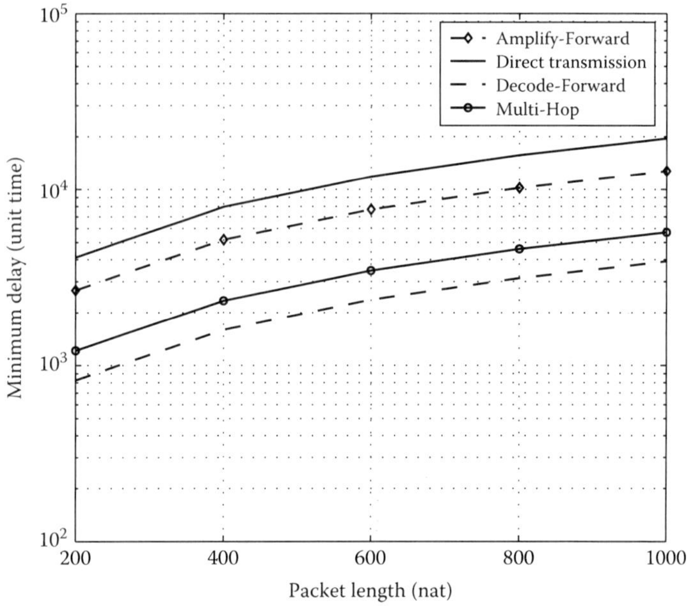

FIGURE 5.1 Minimum link delay versus the original packet in different length (L).

Figure 5.1, all three cooperative protocols achieve much better performance than direct transmission (non-cooperative). Especially, DAF outperforms the two others when SNR remains at a low level. The average delay reductions are 34.19% (AF), 70.57% (MH), and 79.96% (DAF), compared with direct transmission, respectively.

5.3 DELAY ANALYSIS FOR MULTI-HOP SCENARIO

Based on the system model defined in the previous section, now we return to the problem of delay analysis by first characterizing the minimum ETE delay for a multi-hop route. According to (5.7), (5.10), and (5.12), the ETE delay is strongly related to the block error probability and received SINR. A meaningful optimization problem is to minimize the ETE delay in cooperative networks that ensures the ETE error performance satisfied the target level (constraint).

Without loss of generality, let the nodes along the route be denoted as $S \to 1... \to n \to D$. Different from traditional routes, cooperative transmission is used to improve the link quality. However, it is possible that a good helping relay is not available for some pairs of the $n+1$ links of the route. In that case, direct transmission (DT) is used instead of relying on cooperative transmission (CT). Hence the $n+1$ links involved in the route between the source and destination nodes can be categorized into two sets. The first set, defined as $\mathcal{S}_1$, includes the links using only direct transmission, and the other one, defined as $\mathcal{S}_2$, includes all links involved in cooperative transmission. For example, the link between S and 1 in Figure 5.2(a) is included in $\mathcal{S}_1$, whereas in Figure 5.2(b) it is included in $\mathcal{S}_2$. Note that $|\mathcal{S}_1| + |\mathcal{S}_2| = n+1$ since there are only $n+1$ links on the route.

The problem to minimize ETE delay in cooperative networks (e.g., using AF) with the constraint on ETE reliability ϵ can be formulated as

$$\begin{aligned} \min_{\epsilon_{i,j}^{\mathrm{DT}},\epsilon_{i,j}^{\mathrm{CT}}} \quad & \sum_{i,j\in\mathcal{S}_1} D_{i,j}^{\mathrm{DT}} + \sum_{i,j\in\mathcal{S}_2} D_{i,j}^{\mathrm{CT}} \\ \text{s.t.} \quad & 1 - \prod_{i,j\in\mathcal{S}_1} (1-\epsilon_{i,j}^{\mathrm{DT}}) \prod_{i,j\in\mathcal{S}_2} (1-\epsilon_{i,j}^{\mathrm{CT}}) \le \epsilon\,. \end{aligned} \tag{5.13}$$

For small block error probabilities $\epsilon_{i,j}^{\mathrm{DT}} \ll 1$ and $\epsilon_{i,j}^{\mathrm{CT}} \ll 1$, we can have the following approximation:

$$1 - \prod_{i,j\in\mathcal{S}_1} (1-\epsilon_{i,j}^{\mathrm{DT}}) \prod_{i,j\in\mathcal{S}_2} (1-\epsilon_{i,j}^{\mathrm{CT}}) \approx \sum_{i,j\in\mathcal{S}_1} \epsilon_{i,j}^{\mathrm{DT}} + \sum_{i,j\in S_2} \epsilon_{i,j}^{\mathrm{CT}}\,. \tag{5.14}$$

So the optimization problem can be simplified as

$$\begin{aligned} \min_{\epsilon_{i,j}^{\mathrm{DT}},\epsilon_{i,j}^{\mathrm{CT}}} \quad & \sum_{i,j\in\mathcal{S}_1} D_{i,j}^{\mathrm{DT}} + \sum_{i,j\in\mathcal{S}_2} D_{i,j}^{\mathrm{CT}} \\ \text{s.t.} \quad & \sum_{i,j\in S_1} \epsilon_{i,j}^{\mathrm{DT}} + \sum_{i,j\in S_2} \epsilon_{i,j}^{\mathrm{CT}} \le \epsilon\,. \end{aligned} \tag{5.15}$$

By introducing an auxiliary variable z, (5.15) can be written as

$$\begin{aligned} \min_{\epsilon_{i,j}^{\mathrm{DT}},\epsilon_{i,j}^{\mathrm{CT}},z} \quad & \sum_{i,j\in\mathcal{S}_1} D_{i,j}^{\mathrm{DT}} + \sum_{i,j\in\mathcal{S}_2} D_{i,j}^{\mathrm{CT}} \\ \text{s.t.} \quad & \sum_{i,j\in\mathcal{S}_1} \epsilon_{i,j}^{\mathrm{DT}} \le z \\ & \sum_{i,j\in\mathcal{S}_2} \epsilon_{i,j}^{\mathrm{CT}} \le \epsilon - z \\ & 0 \le z \le \epsilon\,. \end{aligned} \tag{5.16}$$

Hence the optimization problem can be solved in two stages. First, we treat z as a constant and solve the following two subproblems separately:

$$\begin{array}{ll} \min\limits_{\epsilon_{i,j}^{\mathrm{DT}}} \sum_{i,j\in\mathcal{S}_1} D_{i,j}^{\mathrm{DT}} & \min\limits_{\epsilon_{i,j}^{\mathrm{CT}}} \sum_{i,j\in\mathcal{S}_2} D_{i,j}^{\mathrm{CT}} \\ \text{s.t.} \sum_{i,j\in\mathcal{S}_1} \epsilon_{i,j}^{\mathrm{DT}} \le z, & \text{s.t.} \sum_{i,j\in\mathcal{S}_2} \epsilon_{i,j}^{\mathrm{CT}} \le \epsilon - z. \end{array} \tag{5.17}$$

This yields the two solutions

$$\sum_{i,j\in\mathcal{S}_1} D_{i,j}^{\mathrm{DT}} = \sum_{i,j\in\mathcal{S}_1} K_{i,j}\left(\rho L - \ln\left(\frac{zK_{i,j}}{\sum_{i,j\in\mathcal{S}_1} K_{i,j}}\right)\right), \tag{5.18}$$

$$\sum_{i,j\in\mathcal{S}_2} D_{i,j}^{\mathrm{CT}} = \sum_{i,j\in\mathcal{S}_2} C_{i,j}\left(\rho L - \ln\left(\frac{C_{i,j}(\epsilon - z)}{\sum_{i,j\in\mathcal{S}_2} C_{i,j}}\right)\right), \tag{5.19}$$

where

$$K_{i,j} = \frac{1}{\rho \ln\left(1 + \frac{\mathrm{SINR}_{i,j}}{1+\rho}\right)} \quad \text{and} \quad C_{i,j} = \frac{2}{\rho \ln\left(1 + \frac{\mathrm{SINR}_{i,j} + f(\mathrm{SINR}_{i,r}, \mathrm{SINR}_{r,j})}{1+\rho}\right)}.$$

The development of (5.18) and (5.19) is provided in Appendix A.3. It is worth noting that both $\sum_{i,j\in\mathcal{S}_1} D_{i,j}^{\mathrm{DT}}$ and $\sum_{i,j\in\mathcal{S}_2} D_{i,j}^{\mathrm{CT}}$ now become functions of the auxiliary variable z.

The second step is to solve the following optimization problem:

$$\begin{array}{lll} \min\limits_{z} & f_z(z) = \sum_{i,j\in\mathcal{S}_1} K_{i,j}\left(\rho L - \ln\left(\frac{zK_{i,j}}{\sum_{i,j\in\mathcal{S}_1} K_{i,j}}\right)\right) \\ & \qquad + \sum_{i,j\in\mathcal{S}_2} C_{i,j}\left(\rho L - \ln\left(\frac{C_{i,j}(\epsilon-z)}{\sum_{i,j\in\mathcal{S}_2} C_{i,j}}\right)\right) \\ \text{s.t.} & 0 \le z \le \epsilon\,. \end{array}$$

Note that $f_z(z)$ is a convex function for $0 \le z \le \epsilon$ since $\frac{d^2 f_z(z)}{d^2 z} > 0$. Hence there is only one minimum value for $0 \le z \le \epsilon$ when $\frac{df_z(z)}{dz} = 0$. We can derive the following:

$$\frac{df_z(z)}{dz} = \sum_{i,j\in\mathcal{S}_1} \frac{K_{i,j}}{z} - \sum_{i,j\in\mathcal{S}_2} \frac{C_{i,j}}{\epsilon - z}. \tag{5.20}$$

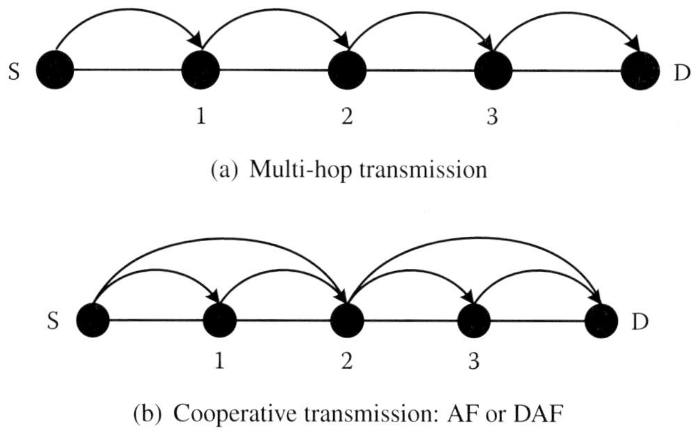

FIGURE 5.2 An example of route selection using different transmission protocols.

Then the optimal distribution of error probability is

$$z^* = \frac{\epsilon \sum_{i,j\in\mathcal{S}_1} K_{i,j}}{\sum_{i,j\in\mathcal{S}_1} K_{i,j} + \sum_{i,j\in\mathcal{S}_2} C_{i,j}}. \tag{5.21}$$

Substituting z^* into $f(z)$, we get the minimum ETE delay. Meanwhile, the optimal error probability for each link can be shown from Appendix A.3 as

$$\epsilon_{i,j}^{\mathrm{DT}} = \frac{z^* K_{i,j}}{\sum_{i,j\in\mathcal{S}_1} K_{i,j}}, \qquad \epsilon_{i,j}^{\mathrm{CT}} = \frac{C_{i,j}(\epsilon - z^*)}{\sum_{i,j\in\mathcal{S}_2} C_{i,j}}. \tag{5.22}$$

It is worth noting that the above results can also be used in both DAF and MH transmissions with a placement of $C_{i,j}$ in each protocol.

To compare the minimum ETE delay achieved by each protocol, we consider a linear network where nodes are assumed to be uniformly distributed between a source-destination pair. L nats of data are transmitted hop by hop from the source to the destination, where only one transmission is allowed along the same route in each time slot. The ETE distance is τ, and there are H equal-length hops in between.

(1) When $\mathcal{S}_2 = 0$

In this scenario, the linear network only employs MH transmission from the source to the destination as shown in Figure 5.2(a). According to (5.22), since $\mathcal{S}_2 = 0$ and $K_{i,j}$ is equal for each hop, the error distribution for each hop is derived as $\epsilon_{i,j}^{\mathrm{DT}} = \frac{\epsilon}{H}$; then we have the minimum ETE delay by using only multi-hop transmission

$$D_{\min}^{\mathrm{MH}} \geq \frac{H(\rho L - \ln(\epsilon/H))}{\rho \ln\left(1 + \frac{\gamma(\tau/H)^{-\alpha}}{1+\rho}\right)}, \tag{5.23}$$

where γ is the transmission power-to-noise ratio. Intuitively, it is not difficult to observe from (5.23) that the delay performance is improved as the total number of hops increases. So, we have the following conclusion.

Theorem 5.1 *For a linear network scenario in low SNR region, the minimum ETE delay for a multiple hops route is proportional to* $\frac{1}{H^{\alpha-1}\tau^{-\alpha}}$, *given* $\tau/H \geq$ *threshold,*[4] *where* H *is the total number of hops and* α *is the path-loss exponent.*

Proof According to (5.23), the rise of total number of hops H will lead to an increase in both the numerator and denominator. However, the denominator will increase with a higher order. Especially, when the transmission power goes to 0, according to Taylor expansion, we have

$$D_{\min}^{\mathrm{MH}} \geq \frac{H(\rho L - \ln(\epsilon/H))}{\rho \ln\left(1 + \frac{\gamma(\tau/H)^{-\alpha}}{1+\rho}\right)} \quad \Rightarrow \quad D_{\min}^{\mathrm{MH}} \geq \frac{H(\rho L - \ln(\epsilon/H))}{\rho\frac{\gamma(\tau/H)^{-\alpha}}{1+\rho}}. \tag{5.24}$$

If we assume the transmitted data L are large enough, then $\rho L \gg \ln(\epsilon/H)$; the lower bound of minimum end-to-end delay can be expressed as

$$D_{\min}^{\mathrm{MH}} \geq \frac{HL}{\frac{\gamma(\tau/H)^{-\alpha}}{1+\rho}} \quad \Rightarrow \quad D_{\min}^{\mathrm{MH}} \geq \frac{L(1+\rho)}{\gamma H^{\alpha-1}\tau^{-\alpha}}, \tag{5.25}$$

which leads to the result. □

Therefore, we can conclude that a route with a large H is preferable for achieving minimum delay.

---

[4] The path-loss model is based on a far-field assumption: the distance is assumed to be much larger than the carrier wavelength. When the distance is on the order of or shorter than the carrier wavelength, the simple path-loss model obviously does not hold anymore as path loss can potentially become path gain. The reason is that near-field electromagnetics now come into play. Therefore, the total number of hops or adjacent distance should not be smaller than a threshold value.

(2) When $\mathcal{S}_1 = 0$

The linear network only prefers cooperative transmission from the source to the destination, either using AF or DAF as shown in Figure 5.2(b). We can derive the optimal

$$C_{i,j}^{\text{AF}} = \frac{2}{\rho \ln\left(1 + \frac{\gamma(2\tau/H)^{-\alpha} + f(\gamma(\tau/H)^{-\alpha}, \gamma(\tau/H)^{-\alpha})}{1+\rho}\right)}$$

to achieve the minimum delay; the error probability for each cooperative link is $\epsilon_{i,j}^{\text{CT}} = \frac{2\epsilon}{H}$. The minimum ETE delay by using only AF is

$$D_{\min}^{\text{AF}} \geq \frac{H(\rho L - \ln(2\epsilon/H))}{\rho \ln\left(1 + \frac{\gamma(2\tau/H)^{-\alpha} + f(\gamma(\tau/H)^{-\alpha}, \gamma(\tau/H)^{-\alpha})}{1+\rho}\right)}. \tag{5.26}$$

For DAF, the minimum ETE delay is

$$D_{\min}^{\text{DAF}} \geq \begin{cases} \dfrac{H(\rho L - \ln(2\epsilon/H))}{\rho \ln\left(1 + \frac{2\gamma(2\tau/H)^{-\alpha}}{1+\rho}\right)}, & \text{if } b > I, \\ \dfrac{H(\rho L - \ln(2\epsilon/H))}{\rho \ln\left(1 + \frac{\gamma(2\tau/H)^{-\alpha} + \gamma(\tau/H)^{-\alpha}}{1+\rho}\right)}, & \text{if } b \leq I, \end{cases} \tag{5.27}$$

where $b = \frac{L}{2N \ln 2}$ bps/unit hertz and the channel capacity $I = \frac{1}{2}\log(1 + \gamma(\tau/H)^{-\alpha})$. It is worth noting that in order to derive a closed-form expression for AF and DAF, here we assume H is an even number; if H is an odd number, it turns to a general case and the result can be directly derived from $f(z)$ and (5.22).

In what follows, we provide a numerical result to illustrate the effect of the criterion of minimizing ETE delay along the path with the constrained ETE reliability using different cooperative protocols. Considering a multi-hop transmission, we assume ETE distance τ between the source and the destination is 30 m, the transmission power-to-noise ratio is 10 dB, the path-loss exponent is set as $\alpha = 3$, $\rho = 0.5$, and the prefixed ETE reliability is $\epsilon = 0.001$. The optimal delay performance of these protocols when sending 200 nats of data from the source is shown in Figure 5.3 as a function of number of hops along the route. As can be seen, the figure confirms Theorem 5.1 that ETE delay decreases as the number of hops in the selected route

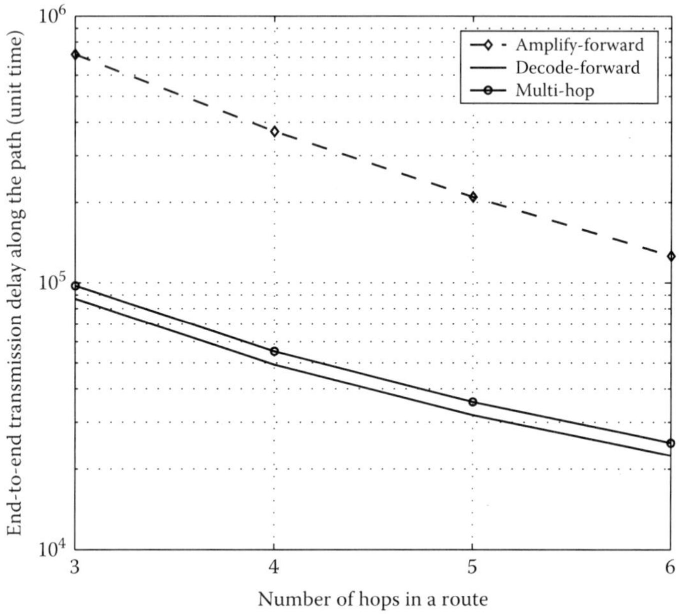

FIGURE 5.3 Total transmission delay along the path versus number of hops in a route.

increases. Furthermore, it also shows that DAF and MH achieve much better performance than AF. This is so because when using AF, noise signal is not removed at the relay as it is amplified and transmitted with the useful signals to the destination, whereas for decode-forward and multi-hop, the relay acts as a second source to transmit the same signals without noise to the destination.

5.4 DELAY ANALYSIS WITH INTERFERENCE SUBTRACTION

In order to further investigate the interference impact on network performance, we consider a more realistic network scenario that allows multi-node transmissions along the same route using the space-time reuse scheme. To tackle the interference, the information of the sets of transmitters in each time slot is needed. In order to simplify the problem and get more meaningful results, here we use a linear network topology in which infinite nodes are regularly placed and each node on the route always has data to send. The distance between adjacent nodes is normalized as 1, and the number of hops between the source and the destination is H. Therefore, given any transmission schedule, each node along the route will experience the same SINR.

We assume all nodes along the route transmit in the same frequency band and employ a regular TDM-schedule of length K-hops so that in time slot t, the nodes $iK + (t \bmod K)$ are allowed to transmit, for $i = ... - 1, 0, 1....$ It is still assumed that the data are transmitted from the source to the destination via multi-hop transmission without queuing delay.

5.4.1 Interference Subtraction

For a multi-hop transmission, the performance of wireless networks can be further improved if prior information available at the receivers can be utilized to achieve perfect interference subtraction. Therefore, we implement the same interference subtraction of Section 4.2, in which multiple transmissions are allowed only when the information in the interfering signal is known at the receiver.

According to the system model, the received SINR at each node is derived as

$$\text{SINR} = \frac{p}{N_0 + \sum_{i=1}^{\infty}(iK+1)^{-\alpha}p + \sum_{i=1}^{\infty}(iK-1)^{-\alpha}p}, \tag{5.28}$$

where p is the transmission power and K is the channel reuse factor. After we implement the interference subtraction, the received SINR can be improved as

$$\text{SINR}' = \frac{p}{N_0 + \sum_{i=1}^{\infty}(iK+1)^{-\alpha}p}. \tag{5.29}$$

For example, we can derive that $\text{SINR}'(K = 2) = \text{SINR}(K = 4)$, which means that employing MUD in wireless networks can potentially increase spatial reuse without losing system performance. Motivated by the fact that prior information available at the receiver can be utilized to achieve perfect interference subtraction by using MUD scheme and therefore invite more simultaneous transmissions along a multi-hop routing, we propose here to further exploit delay performance in cooperative networks by employing the MUD scheme.

Theorem 5.2 *For a regular linear network scenario, the performance gain g, which is defined as the ratio of delay performance under the multi-hop* *scheduling employing the interference subtraction to that without employing the interference subtraction, is bounded by*

$$\frac{T1}{T4} < g < \frac{T2}{T3} < 1,$$

where

$$T1 = \ln\left(1 + \frac{K^{\alpha}(K-1)^{\alpha}}{(1+\rho)(K^{\alpha}+(K-1)^{\alpha})zeta[\alpha]}\right),$$

$$T2 = \ln\left(1 + \frac{K^{\alpha}(K+1)^{\alpha}}{(1+\rho)(K^{\alpha}+(K+1)^{\alpha})zeta[\alpha]}\right),$$

$$T3 = \ln\left(1 + \frac{K^{\alpha}}{(1+\rho)zeta[\alpha]}\right),$$

$$T4 = \ln\left(1 + \frac{(K+1)^{\alpha}}{(1+\rho)zeta[\alpha]}\right),$$

$T3 \neq 0$, $T4 \neq 0$, *K is the channel reuse factor, α is the path-loss exponent,* $zeta[2] = \frac{\pi^2}{6}$, $zeta[3] = 1.202$, *and* $zeta[4] = \frac{\pi^4}{90}$.

Proof See Appendix A.4. □

In general, Theorem 5.2 tells us that the multi-hop scheduling employing the interference subtraction can achieve much better delay performance than that without employing the interference subtraction. For example, for the case where the reuse factor $K = 3$ and the path-loss exponent $\alpha = 3$, the upper bound performance of gap ratio g is 0.41, which means up to 58.67% transmission time can be saved when using the interference subtraction.

5.4.2 End-to-End Delay Analysis

We assume that L nats of data are transmitted in m equal-size packets through a multi-hop route using the space-time reuse scheme. Without considering the additional overheads in each packet, the ETE delay in channel reuse is

$$D_{\text{ETE}} = (H + (m-1)K)D_{ph}, \tag{5.30}$$

where H is the total number of hops between the source and the destination and D_{ph} is the delay per hop. Here, by using the results in Section 5.3, the optimal ETE delay of different cooperative protocols with a constrained ETE reliability ϵ are as follows:

1. *Amplify-and-forward*: The ETE delay is

$$D_{\text{ETE}}^{\text{AF}} \geq \frac{(H+(m-1)K)(\rho\frac{L}{m}-\ln(\frac{2\epsilon}{mH}))}{\rho\ln\left(1+\frac{\text{SINR}_{s,d}+f(\text{SINR}_{s,r},\text{SINR}_{r,d})}{1+\rho}\right)}, \tag{5.31}$$

where

$$\text{SINR}_{s,d} = \frac{p2^{-\alpha}}{N_0+\sum_{i=1}^{\infty}(iK+2)^{-\alpha}p},$$

$$\text{SINR}_{s,r} = \text{SINR}_{r,d} = \frac{p}{N_0+\sum_{i=1}^{\infty}(iK+1)^{-\alpha}p}.$$

2. *Decode-and-forward*: The ETE delay is

$$D_{\text{ETE}}^{\text{DAF}} \geq \begin{cases} \dfrac{(H+(m-1)K)(\rho\frac{L}{m}-\ln(\frac{2\epsilon}{mH}))}{\rho\ln\left(1+\frac{2\text{SINR}_{s,d}}{1+\rho}\right)}, & \text{if } b > I, \\ \dfrac{(H+(m-1)K)\left(\rho\frac{L}{m}-\ln(\frac{2\epsilon}{mH})\right)}{\rho\ln\left(1+\frac{\text{SINR}_{s,d}+\text{SINR}_{r,d}}{1+\rho}\right)}, & \text{if } b \leq I, \end{cases} \tag{5.32}$$

where $b = \frac{L}{2mN\ln 2}$ bps/unit hertz, the channel capacity $I = \frac{1}{2}\log(1+\text{SINR}_{s,r})$, and

$$\text{SINR}_{s,d} = \frac{p2^{-\alpha}}{N_0+\sum_{i=1}^{\infty}(iK+2)^{-\alpha}p},$$

$$\text{SINR}_{s,r} = \text{SINR}_{r,d} = \frac{p}{N_0+\sum_{i=1}^{\infty}(iK+1)^{-\alpha}p}.$$

3. *Multiple-hop*: The ETE delay is

$$D_{\text{ETE}}^{\text{MH}} \geq \frac{(H+(m-1)K)(\rho\frac{L}{m}-\ln(\frac{\epsilon}{mH}))}{\rho\ln\left(1+\frac{\text{SINR}_{s,r}}{1+\rho}\right)}, \tag{5.33}$$

where

$$\text{SINR}_{s,r} = \frac{p}{N_0+\sum_{i=1}^{\infty}(iK+1)^{-\alpha}p}.$$

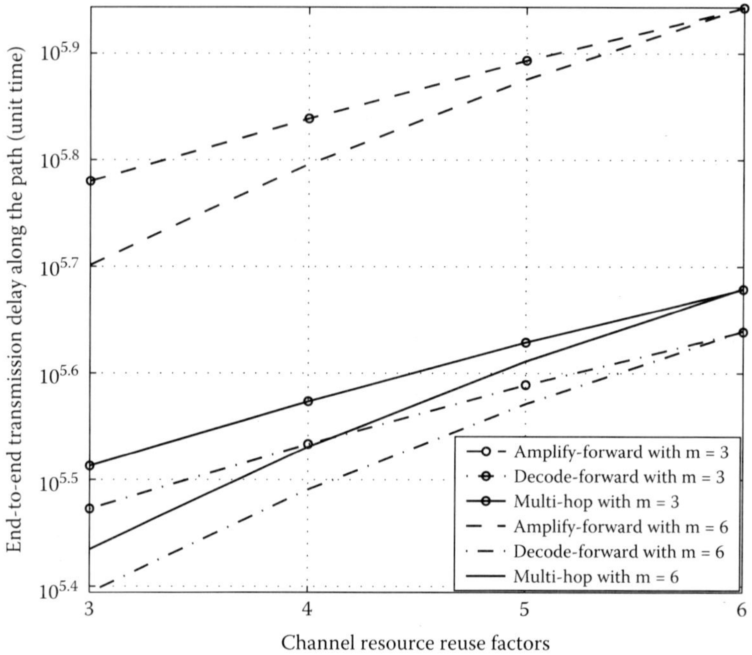

FIGURE 5.4 End-to-end delay performance when using interference subtraction.

It is worth noting that the cooperative protocols are applicable when channel reuse factor $K > 2$, since each cooperative transmission requires two receivers along the route. When $K \leq 2$, only multi-hop transmission is applicable.

The ETE delay performance of different cooperative protocols is shown in Figure 5.4 as a function of channel reuse factor. It is assumed that in total 4×10^4 nats of data are transmitted via a 6-hop route, the transmission power-to-noise ratio is 10 dB, the path-loss exponent is set as $\alpha = 3$, $\rho = 0.9$, and the prefixed ETE reliability is $\epsilon = 0.001$. It is interesting to observe that choosing a larger block number m leads to a better ETE delay performance. In other words, the original data divided into smaller block size is preferable to minimize delay. In addition, as the reuse factor increases, the whole transmission will experience a longer delay. There are two reasons that can explain this. First, according to (5.33), when L is large, the numerator can be simplified as $\rho(KL(\frac{m-1}{m}) + \frac{HL}{m})$, in which K will increase with a higher order than that in the denominator. Second, we are interested in low SNR cases, which means the interference will not dominate the performance even when the reuse factor is small. Furthermore, when the reuse factor approaches its maximum ($K = H$), the ETE delay will not be affected by the block number m as it corresponds to the interference-free scenario.

5.4.3 Throughput Analysis

In order to find more insights on the relation between the delay and any other system performance parameter (e.g., throughput), we are interested in addressing another relevant problem of maximizing the ETE throughput under the same network scenario. The average throughput can be expressed as

$$\lambda = \frac{L}{mKN}, \tag{5.34}$$

where N is the coding length. Under the same system setup, it is clear in Figure 5.5 that the original data divided into larger block size can help achieve larger throughput. To gain some insights, we consider the optimal m and K in low SNR region; for example, using AF, it yields

$$\lambda \le \underbrace{\frac{\frac{L}{m}}{\frac{L}{m} - \frac{1}{\rho}\ln\left(\frac{2\epsilon}{mH}\right)}}_{\mathbf{T}_1} \times \underbrace{\frac{\ln\left(1 + \frac{\frac{p2^{-\alpha}}{N_0+\sum_{i=1}^{\infty}(iK+2)^{-\alpha}p} + f\left(\frac{p}{N_0+\sum_{i=1}^{\infty}(iK+1)^{-\alpha}p}, \frac{p}{N_0+\sum_{i=1}^{\infty}(iK+1)^{-\alpha}p}\right)}{1+\rho}\right)}{K}}_{\mathbf{T}_2}. \tag{5.35}$$

The upper bound performance of throughput is divided into $\mathbf{T}_1$ and $\mathbf{T}_2$, respectively. In order to achieve the maximum value in (5.35), both $\mathbf{T}_1$ and $\mathbf{T}_2$ should be maximized. It is easy to verify that $\mathbf{T}_1$ is maximized when m is as small as possible. In $\mathbf{T}_2$, since the numerator approaches 0 when SNR remains at a low level, the optimal K at the denominator should be the smallest as well.

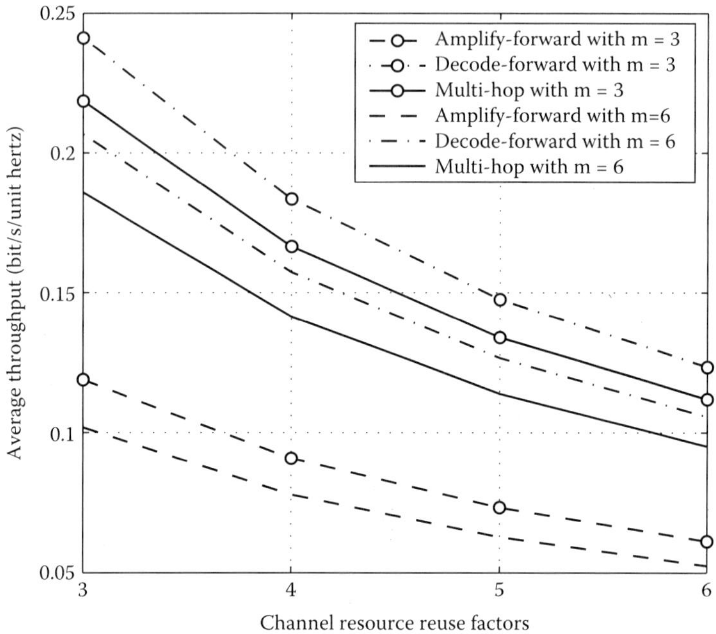

FIGURE 5.5 Average throughput performance when using interference subtraction.

Readers might notice that in fact there is a trade-off between the ETE delay and the network throughput. As block size of the original data decreases, the ETE delay is reduced correspondingly. However, the network throughput is adversely affected by small block size. Under the same assumption of Figure 5.4 and a fixed channel reuse factor $K = 3$, Figures 5.6, 5.7, and 5.8 show such trade-off between the ETE delay and the network throughput of three protocols with and without interference subtraction, respectively. It is worth noting that the channel reuse factor also plays an important role in system performance. Based on the power level that the system selects, the optimal K would be varied by other system parameters.

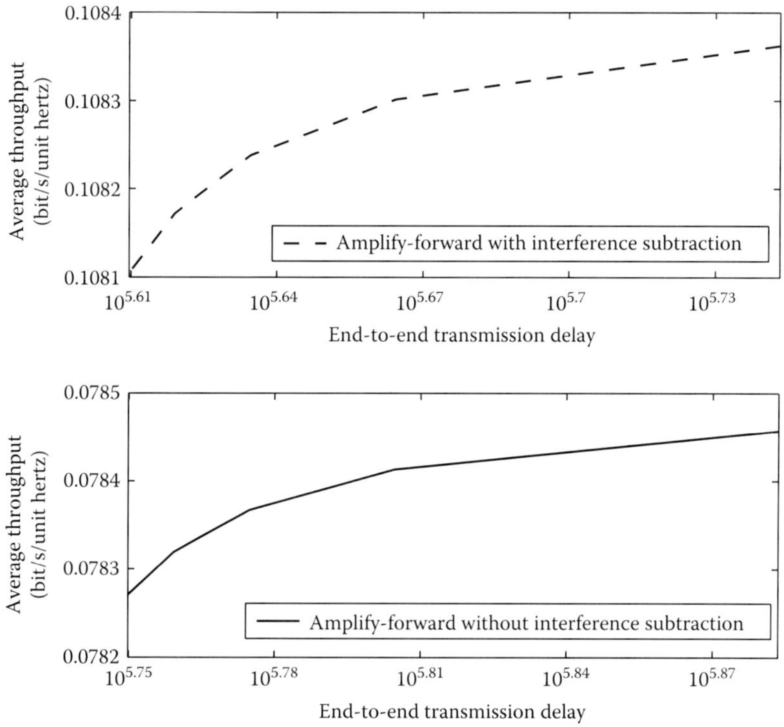

FIGURE 5.6 Average throughput versus end-to-end delay for AF.

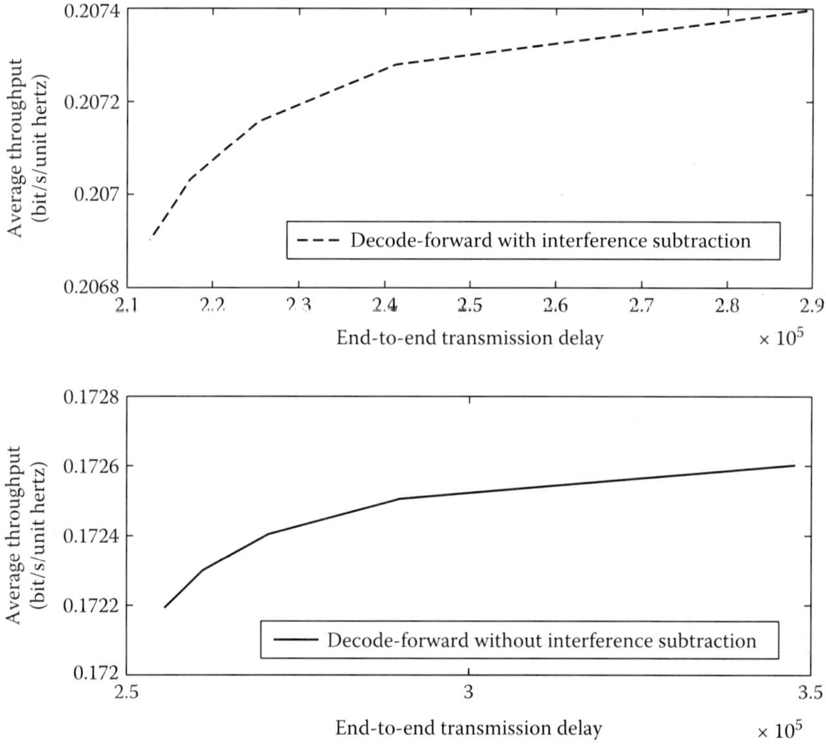

FIGURE 5.7 Average throughput versus end-to-end delay for DAF.

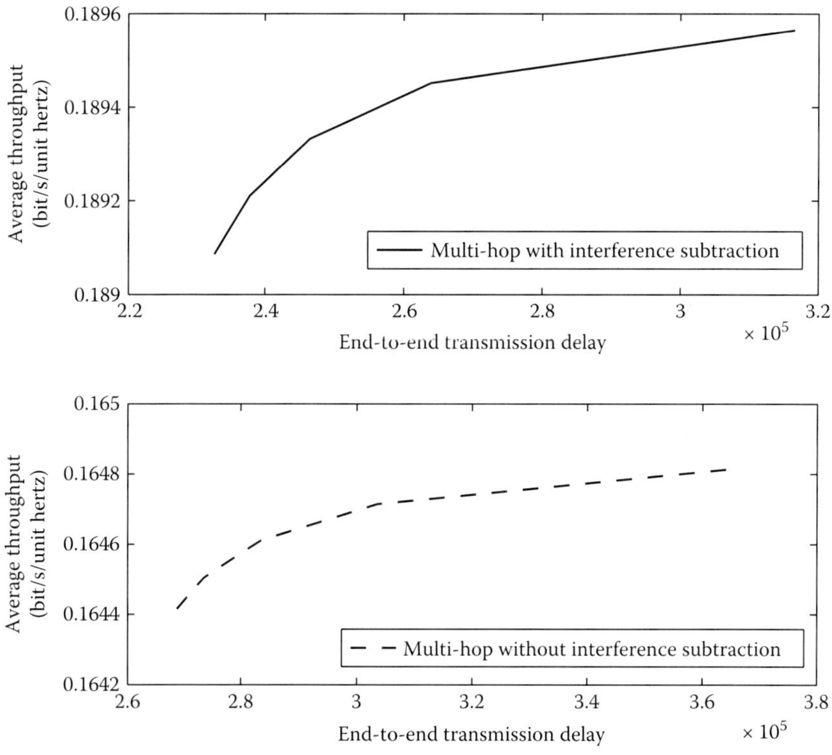

FIGURE 5.8 Average throughput versus end-to-end delay for MH.

II

Cooperative Communication in Single-Hop Scenario

CHAPTER 6

Power Efficiency of Cooperative Transmission

Bongjun Ko, IBM T. J. Watson Research Center, USA
Zhengguo Sheng, University of British Columbia, Canada

6.1 INTRODUCTION

In cooperative communication, the term *cooperation* refers to a node's willingness to sacrifice its own resources (e.g., energy, transmission opportunity) for the benefit of other nodes. It is thus of fundamental importance to understand how much of one's resources must be consumed to reap the benefits of the cooperative communication. Putting it in another way, does cooperative communication require more (or fewer) overall resources than conventional, non-cooperative communication to achieve the same level of wireless link quality? How can we best achieve resource savings when employing cooperative communication? This chapter attempts to answer these fundamental questions.

We start with a single-hop cooperative link and explore a fundamental aspect of cooperative communication (CC): *power consumption*. Since the participation of a wireless device in others' transmissions is critical in cooperative communication, it is of fundamental importance to understand how much energy each participant is required to consume in order to achieve the full benefit of CC. *Our focus is on energy savings of CC*, and as such, we want to know *whether CC can save energy, and if so, under what conditions, and how much*, given a desired quality of the wireless link.

The decode-and-forward (DAF) cooperative protocol considered in this work is similar to that in Yi and Kim [36] and Luo et al. [37], where at least one relay is employed. In contrast, we consider an adaptive version of DAF, which reverts back to direct transmission if the relay cannot decode successfully. More specifically, we investigate power consumption, using at most one relay node as shown in Figure 2.1. As our interest is solely in the power consumption aspects, we assume that solutions to other practical issues in realizing CC are in place (e.g., medium access [38], channel state estimation [39]), which are outside the scope of this chapter.

The following summarizes our contributions and key results:

- We analyze the condition under which CC is preferable to direct transmission and characterize the geometric constraints (which we call the *cooperative region*) on the location of the relay (relative to those of the source and the destination) that lead to lower power consumption. Using the concept of the cooperative region, we provide a probabilistic analysis of the expected energy savings obtained by CC. This is expressed as a function of the node distances, the QoS parameters, and the density of the relays, where the potential relays are assumed to be Poisson distributed. We also show that the average power ratio, defined as the ratio of total transmit power in CC to that of direct transmission, increases as the path-loss exponent or the distance between the source and the destination increases, which indicates that cooperative transmission is more effective in a challenging network environment.

- We derive a closed-form solution for the optimal transmission power required by each source and relay node in DAF cooperative communication under a Rayleigh fading channel model to achieve the given QoS requirements (with targeted data rate and outage probability). Under the optimal power allocation, our analysis shows that the required transmission power of the relay is always smaller than that of the source, a result that lays a foundation to encourage the cooperative behaviors, as this means that the helping party (relay) only needs to spend a relatively small amount of energy compared with the one seeking help from others (source).

- We propose an adaptive cooperation mechanism that will help select appropriate relays for the maximal energy savings of each node in a multi-node environment, and we show that the proposed relay selection can benefit individual nodes from participating in CC. We also study the trade-off between fairness in energy savings and total energy consumption.

6.2 COOPERATIVE REGION

In this section, we establish the conditions under which our cooperative transmission scheme performs better than direct transmission in terms of the power ratio and analyze the geometric properties of the conditions with respect to various parameters.

Given the locations of the source and the destination, we define the *cooperative region* as the geometric region of the location of the relay within which the ratio $\beta = \frac{p_{\text{DAF}}}{p_D}$ is smaller than 1, where p_{DAF} and p_D are transmission power of cooperative and direct transmission, respectively, in (2.12) and (2.6). We define β as a power ratio, so small values of β are preferable. Then the cooperative region is defined by

$$\beta = \frac{p_{\text{DAF}}}{p_D} = \frac{\sqrt{d_{s,r}^{\alpha} + d_{r,d}^{\alpha}}(2^b + 1)\sqrt{2\epsilon^{\text{out}}}}{\sqrt{d_{s,d}^{\alpha}}} < 1\,. \tag{6.1}$$

Further defining a QoS factor $K = 1/((2^b + 1)\sqrt{2\epsilon^{\text{out}}})$, the boundary of the cooperative region is defined by

$$d_{s,r}^{\alpha} + d_{r,d}^{\alpha} = K^2 d_{s,d}^{\alpha}\,. \tag{6.2}$$

Consider the Cartesian coordinate system shown in Figure 6.1, with relay at (x, y), source at $(-\frac{d_{s,d}}{2}, 0)$, and destination at $(\frac{d_{s,d}}{2}, 0)$. Then (6.2) yields

$$\left[\left(x + \frac{d_{s,d}}{2}\right)^2 + y^2\right]^{\frac{\alpha}{2}} + \left[\left(x - \frac{d_{s,d}}{2}\right)^2 + y^2\right]^{\frac{\alpha}{2}} = K^2 d_{s,d}^{\alpha}\,. \tag{6.3}$$

Note that the cooperative region is determined by the QoS factor K, source-destination distance $d_{s,d}$, and path-loss exponent α. In what follows,

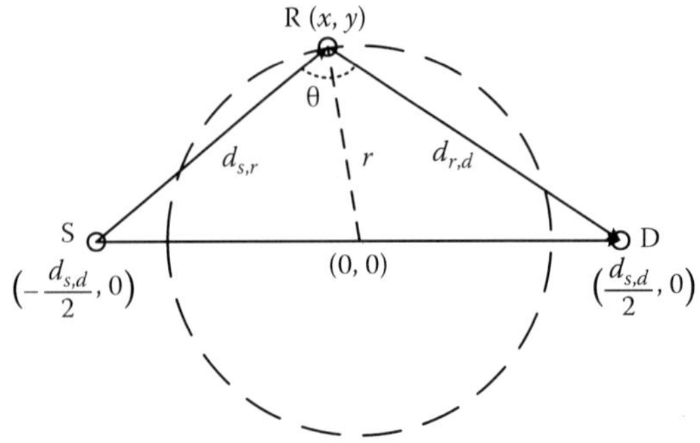

FIGURE 6.1 Geometric analysis for path-loss $\alpha = 2$.

we analyze the characteristics of the cooperative region with regard to these parameters, starting with the special cases of $\alpha = 1$ and $\alpha = 2$.

6.2.1 Path-Loss Exponent α = 1

It is possible to have a path-loss exponent smaller than 2 when there is a waveguide effect, such as in underwater acoustic communications [40] or beamforming. Consider an extreme case $\alpha = 1$ for which the boundary of the cooperative region is

$$d_{s,r} + d_{r,d} = K^2 d_{s,d} \,. \tag{6.4}$$

Thus, the cooperative region is an ellipse in canonical form with foci located at the source and destination and can be described through the canonical equation

$$\frac{x^2}{A^2} + \frac{y^2}{B^2} = 1, \tag{6.5}$$

where $A = \frac{K^2 d_{s,d}}{2}$ and $B = \frac{\sqrt{K^4 - 1} d_{s,d}}{2}$. The area of the cooperative region is $\mathcal{A} = \pi AB$.

6.2.2 Path-Loss Exponent α = 2

According to (6.2), we have

$$d_{s,r}^2 + d_{r,d}^2 = K^2 d_{s,d}^2 \,. \tag{6.6}$$

The cooperative region is a circle, and the foci coincide with the origin (0, 0). With r denoting the distance of the relay from the origin, we have (6.3)

$$x^2 + y^2 = r^2, \quad d_{s,r}^2 + d_{r,d}^2 = 2r^2 + \frac{d_{s,d}^2}{2} = K^2 d_{s,d}^2 \,. \tag{6.7}$$

Hence, the radius of the cooperative region satisfies

$$2\hat{r}^2 + \frac{d_{s,d}^2}{2} = K^2 d_{s,d}^2 \quad \Rightarrow \quad \hat{r} = d_{s,d}\sqrt{\frac{1}{2}\left(K^2 - \frac{1}{2}\right)} \tag{6.8}$$

and the area of the cooperative region is $\mathcal{A} = \pi\hat{r}^2$.

6.2.3 General Path-Loss Exponents

For other path-loss exponents (e.g., $\alpha = 3$ or 4), we can use numerical analysis to characterize the shape of the cooperative region. Motivated by the case of $\alpha = 1$ or 2, it is natural to assume the cooperative region is a general ellipse that can be determined by minor and major radius, A and B. Setting $x = 0$, $y = B$ in (6.3), we can obtain parameter B explicitly as

$$B = d_{s,d}\sqrt{\left(\frac{K^2}{2}\right)^{\frac{2}{\alpha}} - \frac{1}{4}} \,. \tag{6.9}$$

Setting $y = 0$, $x = A$ in (6.3), we can obtain parameter A implicitly via

$$\left|A + \frac{d_{s,d}}{2}\right|^{\alpha} + \left|A - \frac{d_{s,d}}{2}\right|^{\alpha} = K^2 d_{s,d}^{\alpha} \,, \tag{6.10}$$

which can be solved numerically. Then the cooperative region can be defined, approximately, by the ellipse

$$\frac{x^2}{A^2} + \frac{y^2}{B^2} = 1 \,. \tag{6.11}$$

Figure 6.2 illustrates the curves obtained from (6.11) and simulation results for $\alpha = 3$ and 4 when the data rate $b = 2$ bps/Hz, $\epsilon^{\text{out}} = 0.01$, and the source and destination are located at (10 m, 0) and (-10 m, 0), respectively; the two curves are seen to overlap exactly. Moreover, we observe the same in the nu-

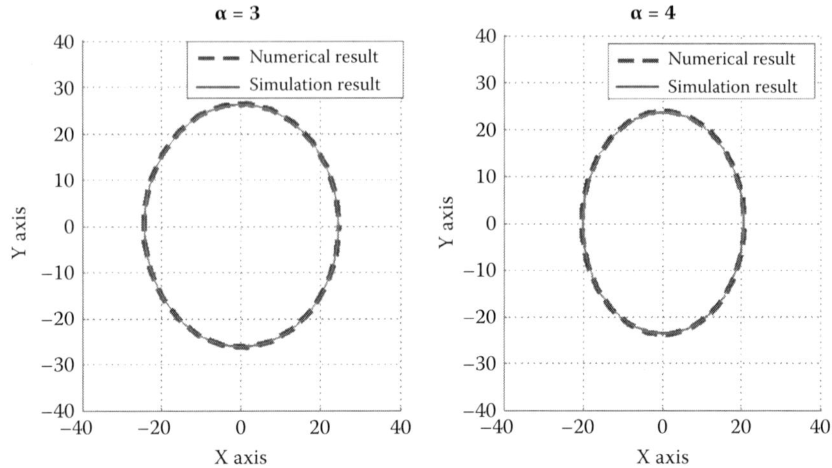

FIGURE 6.2 Cooperative region specified by (6.2) and its ellipse approximation for $\alpha = 3$ and 4.

merical results for different α, b, and ϵ^{out}, indicating that the approximation of the cooperative region by an ellipse is very accurate.

From the above analysis as well as the simulation results shown later in Figure 6.4, we see that the cooperative region, which is a circle for $\alpha = 2$, gets elongated along the x-axis for $\alpha < 2$ and along the y-axis for $\alpha > 2$. Even within the cooperative region, different relays could have different power ratios, and we have the following result on the best relay location.

Lemma 6.1 *For $\alpha > 1$, the best relay location for DAF cooperation is midway between source and destination.*

Proof The best power ratio can be achieved when the left-hand side of (6.3) is minimum, and for any x-coordinate of the relay location, $d_{s,r}$ and $d_{r,d}$ is minimum at $y = 0$. Setting $y = 0$, we can obtain

$$f(x) = \left| x + \frac{d_{s,d}}{2} \right|^{\alpha} + \left| x - \frac{d_{s,d}}{2} \right|^{\alpha}. \tag{6.12}$$

Obtaining the first-order derivative $f'(x) = \alpha(x + \frac{d_{s,d}}{2})^{\alpha-1} - \alpha(\frac{d_{s,d}}{2} - x)^{\alpha-1}$ for $-\frac{d_{s,d}}{2} < x < \frac{d_{s,d}}{2}$, we have $f'(0) = 0$. Moreover, it is not difficult to observe that $f'(x) > 0$ for $x > 0$, and due to symmetry of $f(x)$, we have the similar result $f'(x) < 0$ for $x < 0$. This shows that $f(x)$ monotonically decreases for $x < 0$ and monotonically increases for $x > 0$, and hence $f(x)$ is minimum at $x = 0$. □

Notice that for $\alpha = 1$, $f(x)$ in (6.12) is constant over $-\frac{d_{s,d}}{2} < x < \frac{d_{s,d}}{2}$. The first-order derivative of $f(x)$ is 0, and all points on the line segment between source and destination can achieve the minimum value.

Lemma 6.2 *The minimum K (QoS factor) to guarantee the existence of the cooperation region is $\sqrt{2^{1-\alpha}}$, i.e., $\epsilon^{\text{out}} < 1/[(2^b+1)^2 2^{2-\alpha}]$.*

Proof From Lemma 6.1, the left-hand side of (6.3) gives the minimum when $x = 0$ and $y = 0$, then we have the right-hand side of (6.3) satisfying $K^2 d_{s,d}^{\alpha} \geq 2(\frac{d_{s,d}}{2})^{\alpha}$. Therefore, we can obtain $K \geq \sqrt{2^{1-\alpha}}$. □

Thus, DAF is useful when low outage is required.

Theorem 6.3 *The area of the cooperative region depends on the QoS factor $K = ((2^b+1)\sqrt{2\epsilon^{\text{out}}})^{-1}$, the path-loss exponent α, and transmission distance $d_{s,d}$, and is bounded*[1] *by*

$$\pi\left[\left(\frac{K^2}{2}\right)^{\frac{1}{\alpha}} - \frac{1}{2}\right]^2 d_{s,d}^2 < \mathcal{A}(\alpha) < \pi\left(\frac{K^2}{2}\right)^{\frac{2}{\alpha}} d_{s,d}^2\,. \tag{6.13}$$

Proof From (6.9), we obtain $B < d_{s,d}(\frac{K^2}{2})^{\frac{1}{\alpha}}$. From (6.10), we can obtain $A > d_{s,d}(\frac{K^2}{2})^{\frac{1}{\alpha}} - \frac{d_{s,d}}{2}$. Note that the lower and upper bound are given by a circle with radii A and B, respectively.

The area of the cooperative region given by the ellipse with radii A and B is bounded by $\pi A^2 < \mathcal{A}(\alpha) < \pi B^2$. □

Figure 6.3 shows the area of the cooperative region, obtained via numerical evaluation of (6.2) versus $K^{4/\alpha}$. The linear relationship seen in the curve verifies the theoretical result in Theorem 6.3 and confirms that the elliptical approximation is very accurate.

In essence, the size of the cooperative region increases as the path-loss exponent, targeted data rate, or outage probability decreases. Moreover, a longer transmission distance between the source and destination also indicates an extended opportunity for benefiting from the cooperation when the link condition between the source and the destination is poor.

[1]The lower bound is only valid when $\alpha > 2$.

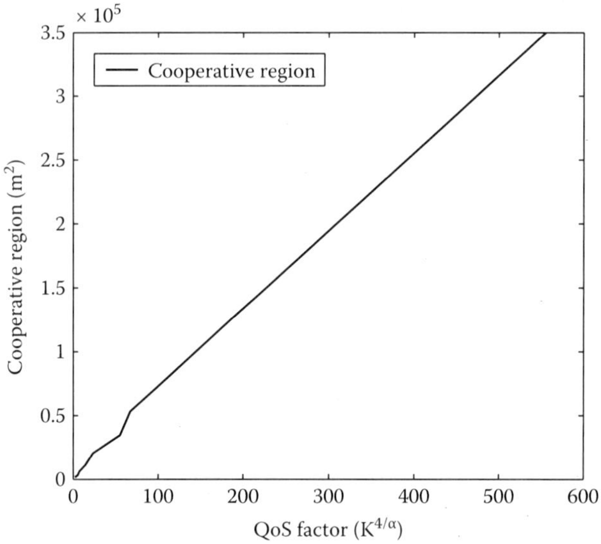

FIGURE 6.3 Area of cooperative region versus QoS factor.

6.2.4 Simulation Result

Figure 6.4 shows the cooperative regions for different path-loss exponents. We assume the data rate $b = 2$ bps/Hz, $\epsilon^{\text{out}} = 0.01$, and the source and destination are located at (10 m, 0) and (−10 m, 0), respectively. The darker the color is, the better the power ratio (lower values of β) can be achieved. It is also clear that as the path-loss exponent increases, the cooperative region becomes smaller.

6.3 AVERAGE POWER RATIO

In this section, we further investigate how much transmission power can be saved by using cooperative transmission and propose a dynamic cooperation scheme. We assume that relay candidates are randomly located in space according to a Poisson point process with density λ. A source-destination pair will choose the best relay node to achieve the minimum total transmission power among all available relay candidates, where the best relay is the one that results in the best power ratio provided in (6.1). A network with a higher density of relay nodes can provide better choices for relay selection.

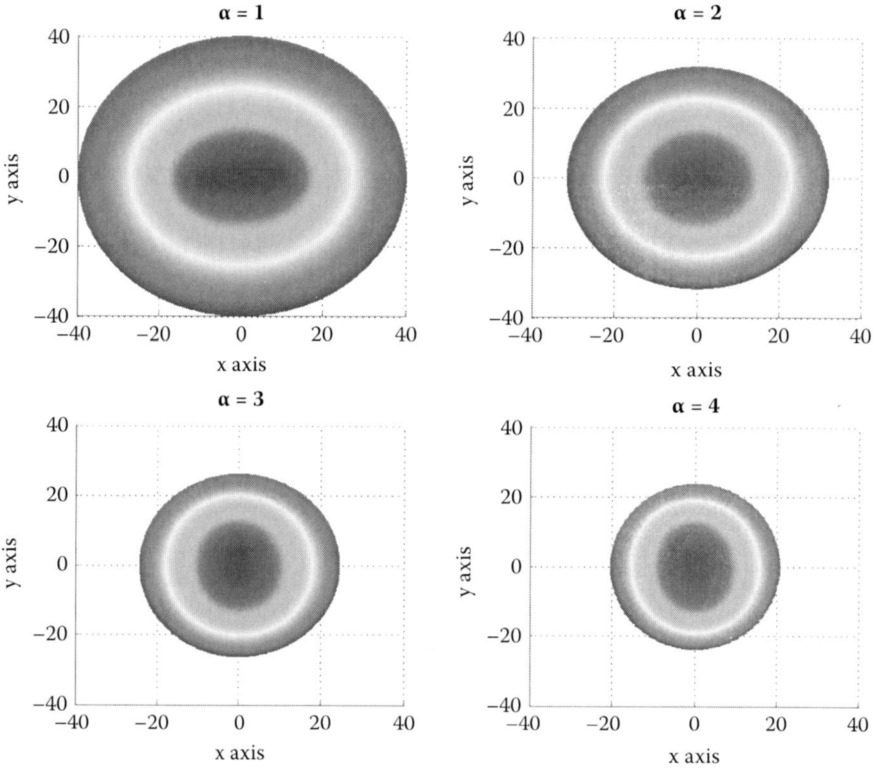

FIGURE 6.4 Cooperative regions for different path-loss exponents.

6.3.1 Average Power Ratio for $\alpha = 2$

When the path-loss exponent $\alpha = 2$, the selected relay to achieve the minimum β will be as close as possible to the origin (0, 0). We let r^* be a random variable of the selected relay distance to the destination and r denote the distance between the closest relay and the destination. The probability distribution function of r is given by

$$\begin{aligned}\Pr[r^* < r] &= 1 - \Pr[r^* \geq r] \\ &= 1 - \Pr[N_r = 0] = 1 - e^{-\lambda\pi r^2},\end{aligned} \tag{6.14}$$

where N_r is the number of relays within distance r from the origin. The probability density function (pdf) of the selected relay distance is

$$f(r) = 2\lambda\pi r e^{-\lambda\pi r^2}, \quad r \geq 0. \tag{6.15}$$

According to (6.1) and (6.7), the expected value of the power ratio is

$$\mathrm{E}\left[\beta\right] = \mathrm{E}\left[\sqrt{\frac{d_{s,r}^{\alpha} + d_{r,d}^{\alpha}}{d_{s,d}^{\alpha} K^2}}\right] = \mathrm{E}\left[\frac{\sqrt{2}}{K}\sqrt{\frac{1}{4} + \frac{r^2}{d_{s,d}^2}}\right], \tag{6.16}$$

where the pdf of the random variable r is given by (6.15). We can have

$$\mu := \mathrm{E}\left[\sqrt{\frac{1}{4} + \frac{r^2}{d_{s,d}^2}}\right] = 2\lambda\pi \int_0^{\infty} \sqrt{\frac{1}{4} + \frac{r^2}{d_{s,d}^2}} r e^{-\lambda\pi r^2} dr. \tag{6.17}$$

Let $y = \frac{1}{4} + \frac{r^2}{d_{s,d}^2}$, then $\frac{2r}{d_{s,d}^2} dr = dy$, $r dr = \frac{d_{s,d}^2}{2} dy$ and $r^2 = d_{s,d}^2(y - \frac{1}{4})$, so that

$$\mu = \lambda\pi d_{s,d}^2 e^{\frac{\lambda\pi d_{s,d}^2}{4}} \int_{\frac{1}{4}}^{\infty} y^{\frac{1}{2}} e^{-\lambda\pi d_{s,d}^2 y} dy. \tag{6.18}$$

Further let $\gamma = \lambda\pi d_{s,d}^2$, and $\gamma y = t$; then recalling the definition of the incomplete upper gamma function

$$\Gamma(u, x) := \int_x^{\infty} e^{-t} t^{u-1} dt,$$

where $u > 0$, we have

$$\mu = \frac{e^{\frac{\lambda\pi d_{s,d}^2}{4}}}{\sqrt{\lambda\pi d_{s,d}^2}}\Gamma\left(\frac{3}{2}, \frac{\lambda\pi d_{s,d}^2}{4}\right), \tag{6.19}$$

which establishes the expectation of power ratio

$$\mathrm{E}\left[\beta\right] = \frac{\sqrt{2}e^{\frac{\lambda\pi d_{s,d}^2}{4}}}{K\sqrt{\lambda\pi d_{s,d}^2}}\Gamma\left(\frac{3}{2}, \frac{\lambda\pi d_{s,d}^2}{4}\right), \tag{6.20}$$

where $\Gamma(\alpha, x) = \int_x^\infty e^{-t}t^{\alpha-1}dt$ is the incomplete gamma function.

Theorem 6.4 *The average power ratio of DAF cooperation relative to direct transmission for $\alpha = 2$ is*

$$\frac{1}{\sqrt{2}K}\sqrt{\frac{\pi}{4\rho}} < E[\beta] < \frac{1}{\sqrt{2}K}\left(\sqrt{\frac{\pi}{4\rho}} + 1\right). \tag{6.21}$$

Proof Let $\rho := \pi\lambda d_{s,d}^2/4$. From the definition of the incomplete gamma function, we have

$$g := e^{\rho}\Gamma\left(\frac{3}{2}, \rho\right) = \int_\rho^\infty t^{\frac{1}{2}}e^{\rho-t}dt\,. \tag{6.22}$$

1. Upper bound:

$$g = \int_0^\infty (\rho+s)^{\frac{1}{2}}e^{-s}ds < \int_0^\infty (\rho^{\frac{1}{2}} + s^{\frac{1}{2}})e^{-s}ds = \rho^{\frac{1}{2}} + \Gamma\left(\frac{3}{2}\right). \tag{6.23}$$

2. Lower bound:

$$g > \int_0^\infty t^{\frac{1}{2}}e^{-t}dt = \Gamma\left(\frac{3}{2}\right) = \frac{\sqrt{\pi}}{2}. \tag{6.24}$$

Using the two bounds in (6.20) leads to (6.21). □

Notice that the parameter $\rho := \pi\lambda d_{s,d}^2/4$ has a nice interpretation as the expected number of relays in a circle with diameter $d_{s,d}$, the source-destination distance. It is worth noting that targeting a smaller outage probability or a longer distance can lead to better power ratio.

6.3.2 General Path-Loss Exponent

The average power ratio for the general case is

$$\mathrm{E}\left[\beta\right] = \mathrm{E}\left[\frac{\sqrt{d_{s,r}^{\alpha} + d_{r,d}^{\alpha}}}{d_{s,d}^{\frac{\alpha}{2}} K}\right]. \tag{6.25}$$

Geometric Lower Bound

We can obtain

$$\mathrm{E_L}\left[\beta\right] = \frac{\sqrt{2}e^{\frac{\lambda\pi d_{s,d}^2}{4}}}{d_{s,d}^{\frac{\alpha}{2}} K (\lambda\pi)^{\frac{\alpha}{4}}} \Gamma\left(\frac{\alpha+4}{4}, \frac{\lambda\pi d_{s,d}^2}{4}\right). \tag{6.26}$$

The mathematical details are provided in Appendix A.5. It is worth noting that when choosing $\alpha = 2$ in (6.26), we can get the same result as (6.20). Therefore, we have the following result.

Theorem 6.5 *The average power ratio of DAF cooperation relative to direct transmission for path-loss exponent α is lower bounded by*

$$E\left[\beta\right] > \frac{\sqrt{2}}{K}\left(\frac{1}{\rho}\right)^{\frac{\alpha}{4}} \Gamma\left(\frac{\alpha+4}{4}\right), \tag{6.27}$$

where $\rho = \pi\lambda d_{s,d}^2/4$ and α is the path-loss exponent.

Proof According to (6.26), we have the lower bound

$$\begin{aligned} \mathrm{E}\left[\beta\right] &= \frac{\sqrt{2}e^{\frac{\lambda\pi d_{s,d}^2}{4}}}{d_{s,d}^{\frac{\alpha}{2}} K (\lambda\pi)^{\frac{\alpha}{4}}} \int_{\frac{\lambda\pi d_{s,d}^2}{4}}^{\infty} t^{\frac{\alpha+4}{4}-1} e^{-t} dt \\ &= \frac{\sqrt{2}}{d_{s,d}^{\frac{\alpha}{2}} K (\lambda\pi)^{\frac{\alpha}{4}}} \int_{\frac{\lambda\pi d_{s,d}^2}{4}}^{\infty} t^{\frac{\alpha+4}{4}-1} e^{\frac{\lambda\pi d_{s,d}^2}{4}-t} dt \\ &> \frac{\sqrt{2}}{d_{s,d}^{\frac{\alpha}{2}} K (\lambda\pi)^{\frac{\alpha}{4}}} \int_{0}^{\infty} t^{\frac{\alpha+4}{4}-1} e^{-t} dt = \frac{\sqrt{2}}{d_{s,d}^{\frac{\alpha}{2}} K (\lambda\pi)^{\frac{\alpha}{4}}} \Gamma\left(\frac{\alpha+4}{4}\right), \end{aligned} \tag{6.28}$$

where $\Gamma(\frac{\alpha+4}{4})$ is a bounded constant factor. □

In essence, Theorem 6.4 and 6.5 tell us that targeting a smaller outage probability, a larger path-loss exponent, or a longer distance can lead to better power ratio, which means that cooperative transmission can better combat a harsher network environment.

Geometric Upper Bound

According to Figure A.1, keeping r as a constant and moving θ to 0, we can obtain

$$\mathrm{E_U}\left[\beta\right] = \frac{\sqrt{\mathrm{E}\left[(\frac{d_{s,d}}{2}+r)^{\alpha}+(\frac{d_{s,d}}{2}-r)^{\alpha}\right]}}{d_{s,k}^{\frac{\alpha}{2}}K}, \tag{6.29}$$

where $\mathrm{E}[(\frac{d_{s,d}}{2}+r)^{\alpha}] = 2\lambda\pi\int_0^{\infty}(\frac{d_{s,d}}{2}+r)^{\alpha}re^{-\lambda\pi r^2}dr$ and $\mathrm{E}[(\frac{d_{s,d}}{2}-r)^{\alpha}] = 2\lambda\pi\int_0^{\infty}(\frac{d_{s,d}}{2}-r)^{\alpha}re^{-\lambda\pi r^2}dr$.

6.3.3 Dynamic Cooperation Scheme

We propose a dynamic cooperation scheme where cooperative transmission is used only if a relay is available within the cooperative region; otherwise, direct transmission is adopted. We compare its performance with unconditional cooperation where cooperative transmission is always adopted regardless of the location of the relay. Let $\hat{r} := d_{s,d}\sqrt{\frac{1}{2}(K^2-\frac{1}{2})}$ be the radius of cooperative region, which is derived from (6.8). We can derive an expression for the mean power ratio for the dynamic cooperation scheme

$$\mathrm{E}\left[\beta\right] = \hat{\mathrm{E}}\left[\beta\right]\Pr[N_{\hat{r}}>0] + 1\cdot\Pr[N_{\hat{r}}=0]\,. \tag{6.30}$$

From (6.14) we have $\Pr[N_{\hat{r}}=0] = e^{-\pi\lambda\hat{r}^2} = e^{-\delta}$, where $\delta := \pi\lambda\hat{r}^2$. The expected power ratio $\hat{\mathrm{E}}[\beta]$ is when the relay is available within the cooperative region and can be derived similar to (6.20) as

$$\begin{aligned}\hat{\mathrm{E}}[\beta] &= \frac{2\sqrt{2}\lambda\pi}{K}\int_0^{\hat{r}}\sqrt{\frac{1}{4}+\frac{r^2}{d_{s,d}^2}}\,re^{-\lambda\pi r^2}dr \\ &= \frac{1}{\sqrt{2}K}\frac{e^{\rho+\delta}}{\sqrt{\rho+\delta}}\Gamma\left(\frac{3}{2},\rho+\delta\right) - \frac{1}{\sqrt{2}K}\frac{e^{\rho}}{\sqrt{\rho}}\Gamma\left(\frac{3}{2},\rho\right),\end{aligned} \tag{6.31}$$

where $\rho = \pi\lambda d_{s,d}^2/4$ is defined earlier, and $\delta := \pi\lambda\hat{r}^2$.

Note that the scheme requires knowledge of the relay locations.

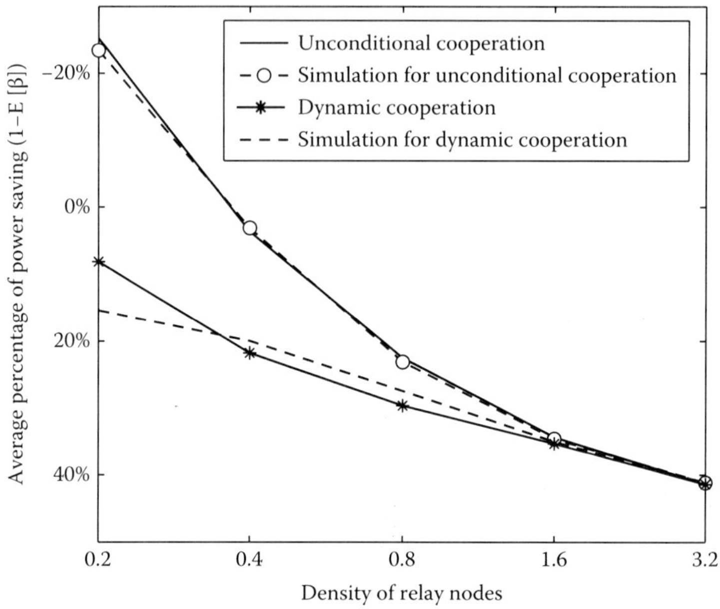

FIGURE 6.5 Average power savings for $\alpha = 2$.

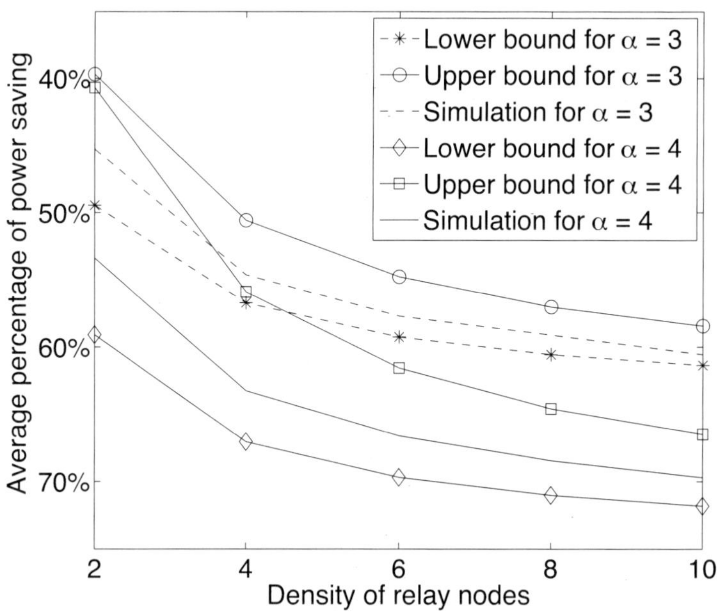

FIGURE 6.6 Average power savings of unconditional cooperation for $\alpha = 3$ and $\alpha = 4$.

6.3.4 Simulation Result

Figure 6.5 shows the performance of the dynamic cooperation scheme. The dynamic cooperation scheme can always guarantee better performance even when the node density is low. Moreover, theoretical results are seen to be very close to the simulation results. Figure 6.6 shows the average power savings $(1 - \mathrm{E}[\beta])$ for other path-loss exponents; it shows that the theoretical bounds in (6.26) and (6.29) well define the behavior of β for general path-loss cases, and furthermore we can observe that a larger path-loss exponent can lead to better power savings.

CHAPTER 7

Optimal Power Allocation of Cooperative Transmission

Zhengguo Sheng, University of British Columbia, Canada
Bongjun Ko, IBM T. J. Watson Research Center, USA

7.1 INTRODUCTION

In previous chapters, we introduce the concept of cooperative region and average power ratio and prove that cooperative communication is effective in enhancing performance of wireless networks. However, cooperative communication techniques typically assume uniform transmission power at every node, including relays. Recall from Chapter 2, we assume that both source and relay employ an identical transmission power p. In this chapter, we propose a scheme to optimize the transmission powers for the source and relay nodes as a means to reduce the total power consumption p_{DAF}, while maintaining the required QoS. Specifically, we propose an optimal power allocation method for the DAF wireless cooperative networks and investigate its fundamental characteristics in terms of power ratio.

7.1.1 Problem Formulation

We consider the same cooperative link in Figure 2.1. Assuming that p and q are the source and relay power, respectively, and the relay node can perform perfect decoding when the received SNR exceeds a threshold, the channel capacity of this cooperative link can be shown as

$$I_{s,d} = \begin{cases} \frac{1}{2}\log(1+2p|\mathrm{a}_{s,d}|^2), & |\mathrm{a}_{s,r}|^2 < f(p)\,, \\ \frac{1}{2}\log(1+p|\mathrm{a}_{s,d}|^2 + q|\mathrm{a}_{r,d}|^2), & |\mathrm{a}_{s,r}|^2 \geq f(p)\,. \end{cases} \tag{7.1}$$

Therefore, the outage event is given by $I_{s,d} < b$, and the outage probability becomes

$$\begin{aligned} \epsilon^{\text{out}} &= \Pr[I_{s,d} < b] \\ &= \Pr[|\mathrm{a}_{s,r}|^2 < f(p)]\Pr[2|\mathrm{a}_{s,d}|^2 < f(p)] \\ &\quad + \Pr[|\mathrm{a}_{s,r}|^2 \geq f(p)]\Pr\left[|\mathrm{a}_{s,d}|^2 + \left|\sqrt{\frac{q}{p}}\mathrm{a}_{r,d}\right|^2 < f(p)\right]. \end{aligned} \tag{7.2}$$

By computing the limit, we obtain from (7.2)

$$\begin{aligned} \frac{1}{f^2(p)}\epsilon^{\text{out}} &= \underbrace{\frac{1}{f(p)}\Pr\left[|\mathrm{a}_{s,r}|^2 < f(p)\right]}_{\mathbf{T1}}\underbrace{\frac{1}{f(p)}\Pr\left[2|\mathrm{a}_{s,d}|^2 < f(p)\right]}_{\mathbf{T2}} \\ &\quad + \underbrace{\Pr[|\mathrm{a}_{s,r}|^2 \geq f(p)]}_{\mathbf{T3}}\underbrace{\frac{1}{f^2(p)}\Pr\left[|\mathrm{a}_{s,d}|^2 + \left|\sqrt{\frac{q}{p}}\mathrm{a}_{r,d}\right|^2 < f(p)\right]}_{\mathbf{T4}} \end{aligned} \tag{7.3}$$

where $\mathbf{T1} \approx d_{s,r}^{\alpha}$, $\mathbf{T2} \approx d_{s,d}^{\alpha}/2$, $\mathbf{T3} \approx 1$, $\mathbf{T4} \approx \frac{p}{q}d_{s,d}^{\alpha}d_{r,d}^{\alpha}/2$. Since $f(p) = \frac{2^{2b}-1}{p}$, we obtain a closed-form expression for the outage probability between the source and the destination using cooperative transmission

$$\epsilon_C^{\text{out}} = \frac{1}{2}d_{s,d}^{\alpha}\left(d_{s,r}^{\alpha} + \frac{p}{q}d_{r,d}^{\alpha}\right)\frac{(2^{2b}-1)^2}{p^2}\,. \tag{7.4}$$

A meaningful optimization problem is to minimize the total transmission power consumption of a cooperative link given that a target QoS is satisfied and can be formulated as

$$\begin{aligned} \min \quad & p+q \\ \text{s.t.} \quad & \epsilon_C^{\text{out}}(p,q) \le \eta, \end{aligned} \tag{7.5}$$

where

- p and q denote the source and relay power, respectively, and $\epsilon_C^{\text{out}}(p,q)$ is the outage probability defined by (7.4),
- A QoS requirement is decided by the target outage probability η and transmission data rate b.

Theorem 7.1 *The optimal transmission power to minimize the total power consumption of DAF cooperation, given that a target QoS is satisfied, is given by*

$$p^* = \sqrt{\frac{A+2B}{2} + \frac{\sqrt{A^2+8AB}}{2}}, \quad q^* = \frac{Ap^*}{p^{*2}-B}, \tag{7.6}$$

where $A = (\mu d_{s,d}^{\alpha} d_{r,d}^{\alpha})/2\eta$, $B = (\mu d_{s,d}^{\alpha} d_{s,r}^{\alpha})/2\eta$, $\mu = (2^{2b}-1)^2$, *and* η *is the outage constraint.*

Proof See Appendix A.6. □

Lemma 7.2 *The optimal relay power* q^* *is always smaller than the optimal source power* p^* *with*

$$p^* > q^* . \tag{7.7}$$

The result follows (7.6) *and has* $q* = 2p^*/[1+\sqrt{1+8d_{s,r}^{\alpha}/d_{r,d}^{\alpha}}]$, *which is always smaller than* p^*. *In general, we find that the optimal DAF cooperation saves the relay power as it moves closer to the destination.*

Lemma 7.3 *The total transmission power of the optimal DAF cooperation is bounded by*[1]

$$p_{\text{con}} < p^* + q^* < 2p_{\text{con}} . \tag{7.8}$$

Proof According to Lemma 7.2, we have the following bound performance on total power consumption:

[1] We refer to the DAF cooperation with identical power assumption ($p = q$) as the conventional cooperation and denote it as p_{con}.

1. Upper bound: when the relay approaches the source node or goes to infinite, we derive $p = q = p_{\text{con}}$, and the optimal cooperation uses the same amount of power as the conventional cooperation.

2. Lower bound: when the relay approaches the destination, we have the optimal source power $p = p_{\text{con}}$, and the performance gain can reach its maximum with the relay power down to 0. □

In essence, the optimal cooperation can help reduce the total transmission power, which will be further analyzed in the following.

7.1.2 Analysis of Optimal DAF Cooperation

In this section, we analyze the optimal DAF cooperative transmission in detail and compare its performance with that of direct transmission. According to (7.6), we have the minimum total power consumption

$$p_{\text{DAF}} = p^* + q^* . \tag{7.9}$$

Cooperative Region for Optimal DAF

We establish the conditions under which our optimal cooperative scheme performs better than direct transmission in terms of power ratio and analyze the geometric properties of the conditions with respect to various parameters.

Using the same definition of (6.1), the cooperative region of optimal DAF is

$$\beta := \frac{p^* + q^*}{p_D} = \frac{\sqrt{\frac{m+1}{4}}\left(\sqrt{d_{s,r}^\alpha} + \frac{d_{r,d}^\alpha}{m\sqrt{d_{s,r}^\alpha}}\right)}{K\sqrt{d_{s,d}^\alpha}} < 1, \tag{7.10}$$

where $K = ((2^b + 1)\sqrt{2\epsilon^{\text{out}}})^{-1}$ is the QoS factor, $m = (\gamma + \sqrt{\gamma^2 + 8\gamma})/2$, and $\gamma = d_{r,d}^\alpha / d_{s,r}^\alpha$.

Theorem 7.4 *The area of the cooperative region depends on the QoS factor K, the path-loss exponent α, and transmission distance $d_{s,d}$, and is bounded*[2] *by*

$$\pi\left[\left(\frac{K^2}{2}\right)^{\frac{1}{\alpha}} - \frac{1}{2}\right]^2 d_{s,d}^2 < \mathcal{A}(\alpha) < \pi(2K^2)^{\frac{2}{\alpha}} d_{s,d}^2 \tag{7.11}$$

[2]The lower bound is only valid when $\alpha > 2$.

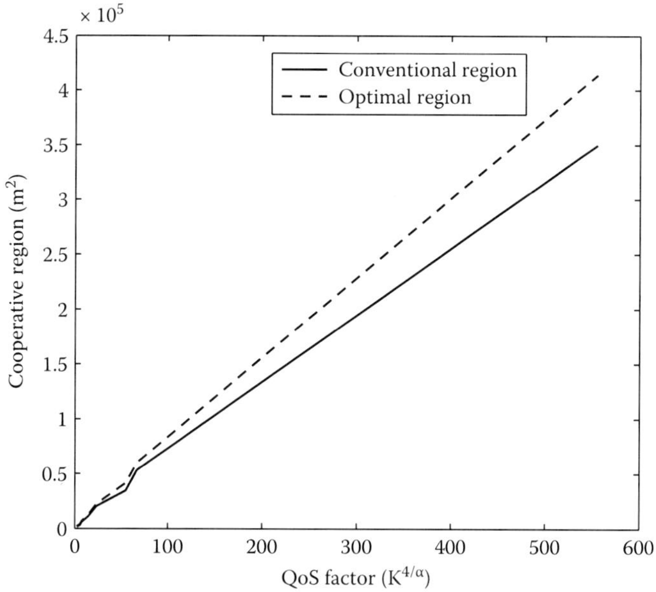

FIGURE 7.1 Comparison of regions size versus scaling factor.

The result is analogue to Theorem 6.3, and according to Lemma 7.3, it can be derived from (6.13). Figure 7.1 verifies the theoretical result in Theorem 7.4 and also shows that the optimal cooperation can achieve a larger region than the conventional cooperation, which indicates an extended opportunity for benefiting from cooperative transmission.

Average Power Ratio of Optimal DAF

In this section, we further investigate how much transmission power can be saved by using optimal cooperation. We still assume that relay candidates are randomly located in space according to a Poisson point process with density λ. A source-destination pair located at $(-\frac{d_{s,d}}{2}, 0)$ and $(\frac{d_{s,d}}{2}, 0)$, respectively, will choose the best relay node to achieve the minimum total transmission power among all available relay candidates, where the best relay is the one that results in the best power ratio defined in (7.10). A network with a higher density of relay nodes can provide better choices for relay selection.

Lemma 7.5 *For $\alpha = 2$, the best relay location that minimizes β for the optimal DAF cooperation is at the destination.*

Proof From (7.10), we can obtain the ratio for $\alpha = 2$

$$\beta = \frac{1}{K}\sqrt{\frac{m+1}{4}}\left(1 + \left(\sqrt{\frac{\gamma}{m^2}} - 1\right)\frac{d_{r,d}}{d_{s,d}}\right),$$

where $\gamma = d_{r,d}^{\alpha}/d_{s,r}^{\alpha}, m = (\gamma + \sqrt{\gamma^2 + 8\gamma})/2$. Since $\beta \geq 0$, it is easy to observe that the minimum value can be obtained as $\frac{1}{2K}$ when $d_{r,d}/d_{s,d} = 0$. □

So the selected relay to achieve the minimum β will be as close as possible to the destination. We note that relays with the same distance r to the destination may not lead to the same β, since the source-to-relay distances may be different, and hence the optimal p^*. But we can use the probability distribution function (6.15) to bound E $[\beta]$ as follows.

Theorem 7.6 *The average power ratio of the optimal DAF cooperation relative to direct transmission for* $\alpha = 2$ *is lower bounded by*

$$E[\beta] > \frac{1 - e^{-\rho}}{2K} + \frac{\sqrt{2}}{K}\int_{\frac{d_{s,d}}{2}}^{\infty}\sqrt{\frac{1}{4} + \frac{\left(r - \frac{d_{s,d}}{2}\right)^2}{d_{s,d}^2}}\,f(r)dr, \qquad (7.12)$$

where $K = ((2^b + 1)\sqrt{2\epsilon^{\text{out}}})^{-1}$, $\rho = \pi\lambda d_{s,d}^2/4$.

Proof See Appendix A.7. □

7.1.3 Simulation Result

In this section, we provide the numerical and simulation results for the optimal DAF cooperation. We first evaluate the power consumption of CC for a single source-destination pair. Here we set the QoS constraints of the bit rate $b = 1$ bps/Hz, and the target outage probability $\epsilon^{\text{out}} = 0.01$, and the source and destination node are placed at the coordinates (10 m, 0) and (−10 m, 0), respectively, in a two-dimensional plane.

Figure 7.2 shows the numerical results for the individual powers of the source and the relay in optimal CC as the location of the relay is varied along the line between the source and the destination. Here the x-axis represents the relative location of the relay with regard to those of the source and destination, and the y-axis is the transmit power in dB. It can be seen that the

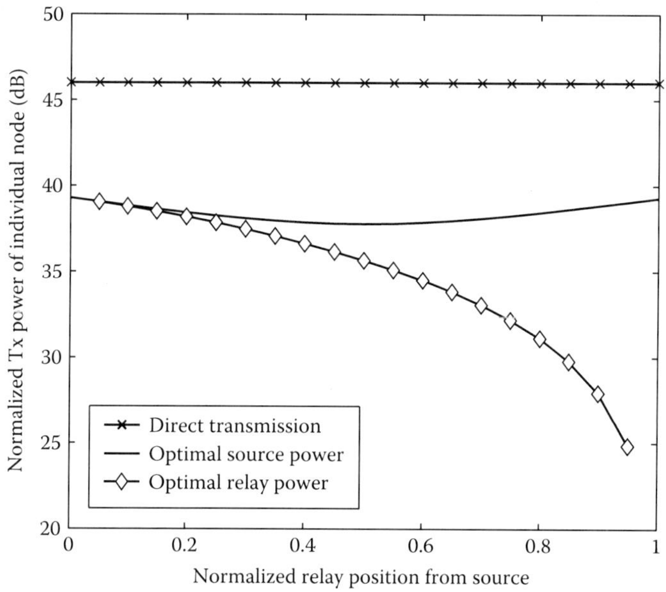

FIGURE 7.2 Individual transmission (Tx) power behaviors for $\alpha = 2$. Normalized relay position from the source is the ratio of distance between source and relay to the distance between source and destination: 0 if relay is close to the source, 1 if relay is close to the destination.

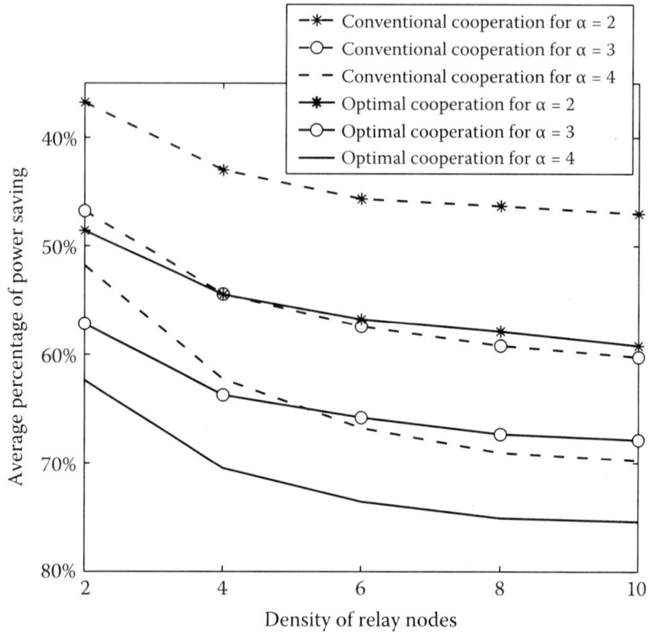

FIGURE 7.3 Average power savings of the optimal cooperation.

source can reduce at least 7 dB of its power compared to the direct transmission. Moreover, the relay's power is always smaller than the source's and monotonically decreases as its location gets closer to the destination.

In Figure 7.3, we plot the average percentage of power savings $(1-\mathrm{E}[\beta])$ of CC for a different pass-loss exponent (α) as we vary the density of the potential relays in the x-axis. To show the power savings of the optimal CC, we also include that of conventional CC (with $p = q$ constraints). The results are averaged over simulating 100 packet transmissions, and the relay with the smallest $p^* + q^*$ is used. The results are consistent with what our analysis predicts: Power savings improve as more relays are available and with larger path-loss exponents (i.e., harsher path-loss). Also, the average improvement of the optimal cooperation over the conventional cooperation can be larger than 20%.

From the above result, the conclusion from the conventional cooperation that targeting a smaller outage probability, a longer distance, or a larger path-loss exponent can lead to a better power ratio is still valid for the optimal

DAF cooperation. Moreover, the optimal cooperation can achieve a much better power ratio.

7.2 ENERGY-EFFICIENT RELAY SELECTION FOR DAF

In the previous sections, we investigate, given a single source and a destination, how much power the cooperative transmission can save for the source and the relay compared to the direct transmission. In this section, we consider a more general network setting where multiple nodes coexist and cooperate with each other by acting as relays for each other's transmissions.

Our interest in this section is to find a set of rules that determine which node to select as the relay for the maximal power savings of each node in this multi-node environment. It is worth noting that relay selection affects the overall energy transmission, since the optimal DAF power depends upon the location of the selected relay and the channel conditions. When multiple relays are available, we expect the overall energy consumption to decrease.

More specifically, our setup consists of a set of nodes $N = \{1, ..., n\}$, where each node $i \in N$ transmits a number of packets over time, each time with some arbitrary destination node in the network. For simplicity, we assume all packets have the same constant length with the same QoS constraints, though it is straightforward to derive relay selection rules in a more general setup. We also assume that time is divided into discrete time slots and that TDMA is used to provide collision-free transmissions from the sources and the relays.[3]

We denote by $p_{i,j}(t)$ and $q_{i,j}(t)$ the transmit power of a source node i and a relay node j, respectively, when i would use cooperative transmission with j as the relay to some destination at time t. We assume that the source and the relay use the optimal transmission powers given by (7.6) for each packet transmission. When node i uses direct transmission at time t, we denote its transmit power as $p^D(t)$. The *energy consumption* of a node $E_i(t_1 : t_2)$ during a time interval $[t_1 : t_2]$ is the sum of node i's transmit power either as a source or a relay over all $t \in [t_1, t_2]$ (we assume a node consumes zero-power at t if it is neither a source nor a relay at t).

---

[3] It is noteworthy, however, that the specific choice of multi-access scheme for cooperative communication is largely orthogonal to our problem of which relay to be selected for energy savings. We discuss an approach to integrating our relay selection rule into a known distributed random access scheme for cooperative communication in Section 7.2.2.

We use $\mathcal{R}_i(t)$ to denote the set of the nodes (except node i) within i's cooperative region (defined in Section 6.2) for source node i's transmission to its destination at time t, i.e., $\mathcal{R}_i(t) = \{j \in N - \{i\} | p_{i,j}(t) + q_{i,j}(t) < p^D(t)\}$.

A *relay-selection rule* is one that assigns each source i's transmission at each time t to some relay, denoted by $r_i(t)$. If no relay is selected at time t for node i, we write $r_i(t) = null$. The goal is to design relay selection rules that can achieve the maximal energy savings due to cooperative transmissions. To represent how much energy savings the cooperative transmission can yield in comparison to direct transmission, we begin by introducing the notion of the "payoffs" of the nodes.

The *payoff function*, $u_i(t)$, of node i at time t is defined as

$$u_i(t) = \begin{cases} p_i^D(t) - p_{i,j}(t) & \text{if } \exists j \text{ s.t., } r_i(t) = j\,, \\ -q_{j,i}(t) & \text{if } i = r_j(t) \text{ for some source } j\,, \\ 0 & \text{otherwise}\,. \end{cases}$$

The above represents how much energy a node i *locally* saves (or loses) compared to direct transmission at time t, where $p_i^D(t) - p_{i,j}(t)$ denotes the power saved from i's cooperative transmission using some relay j at time t, and $-q_{j,i}(t)$ the power spent in i's transmission as a relay for some other node j at time t. In all other cases (if i does not transmit either as a source or a relay, or if i uses direct transmission), the payoff is 0. The initial $u_i(t)$ can be any arbitrary value, but for simplicity, we assume $u_i(t) = 0$ for all $i \in N$. Then the *cumulative payoff* over a time interval $[t_1 : t_2]$ is defined as $u_i(t_1 : t_2) = \sum_{\tau=t_1}^{t_2} u_i(\tau)$, which represents the overall energy savings of a node during the time interval.

7.2.1 Relay Selection Rules

Our first relay-selection rule makes use of the result in previous section in a straightforward manner.

Min-Total-Power Relay Selection

A relay is selected for source i at time t such that

$$r_i(t) = \arg \min_{j \in \mathcal{R}_i(t)} \{p_{i,j}(t) + q_{i,j}(t)\}\,.$$

In other words, for each packet from node i, a relay j is selected that minimizes $p_{i,j}(t)+q_{i,j}(t)$ among those in i's cooperative region at t. If $\mathcal{R}_i(t) = \emptyset$, $r_i(t) = null$.

Note that the Min-Total-Power selection rule is myopic in nature since the selection is based only on the projected power consumptions of itself and other potential relay nodes for the upcoming transmission at each t, but not on the past energy consumptions of itself or other nodes. However, it is easy to see that, though simple, the Min-Total-Power rule is optimal (among all relay selection rules) in the sense that it minimizes the total energy consumption of the network, $\sum_{i \in N} E_i(t_1 : t_2)$ for any time interval $[t_1, t_2]$, and hence maximizes the aggregate cumulative payoffs $\sum_{i \in N} u_i(t_1 : t_2)$ of all nodes.

Theorem 7.7 *For any time interval of* $[t_1, t_2]$, *the total energy consumption of the network* $\sum_{i \in N} E_i(T)$ *is minimized if each* i *is assigned a relay node at each time by the Min-Total-Power rule.*

Proof Since we are only interested in the total energy consumption, we can schedule the whole transmission into several rounds and each node can only transmit no more than one packet in each round. Since any assignment r is injective in each round, for any two nodes i and k, $S_i \cap S_k = \emptyset$, and $\cup_{i \in N} S_i \subseteq N$, where S_i is a set of source nodes whose relay is i. Therefore, the total energy consumption in each round $\sum_{i \in N} E_i = \sum_{i \in N}(p_{i,r_i} + \sum_{j \in S_i} q_{j,i})$ can be rewritten as $\sum_{i \in N} p_{i,r_i} + \sum_{i \in N} \sum_{j \in S_i} q_{j,i} = \sum_{i \in N} p_{i,r_i} + \sum_{j \in N} q_{j,r_j} = \sum_{i \in N}(p_{i,r_i} + q_{i,r_i})$, which is minimized if each individual term $p_{i,r_i} + q_{i,r_i}$ is minimum. □

In other words, the relay assignments that yield the minimum total energy consumption can be simply obtained by having each source node select a relay node such that the combined transmission power for the source and the relay is minimum.

From the individual nodes' perspective, however, the relay selection by the Min-Total-Power rule can lead to the situation that some nodes end up with higher energy consumption than would be the case when all nodes employ direct transmission. This is especially true if some unfortunate nodes are heavily selected as relays and hence consume more energy in relaying than was saved from its own transmission as a source. We now consider how to handle such unfairness issues in CC.

The main idea of the adaptive relay selection is to let each node act as a relay only when it has saved more energy than it has lost from cooperative transmission in the past. For this, a binary decision variable $C_i(t)$ is maintained for each node i and updated at each time t (hence the term *adaptive*) such that

$$C_i(t) = \begin{cases} 1 & \text{if } u_i(0:t-1) \geq 0\,, \\ 0 & \text{if } u_i(0:t-1) < 0\,. \end{cases}$$

This $C_i(t)$ value is used in the decision as to whether node i can act as a relay for other nodes (when $C_i(t) = 1$, i.e., in "cooperative" mode) or i should not be selected as a relay for any other node (when $C_i(t) = 0$).

Adaptive Relay Selection

A relay is selected for source i at time t such that

$$r_i(t) = \arg \min_{j \in \mathcal{R}_i(t), C_j(t)=1} \{p_{i,j}(t) + q_{i,j}(t)\}\,.$$

In other words, a relay j is selected for the i's transmission at time t that minimizes $p_{i,j}(t) + q_{i,j}(t)$ among the nodes whose cumulative payoffs are positive or zero.[4] Thus, a node whose cumulative payoff is negative will cease to act as a relay and will be potentially available as a relay when its payoff becomes positive. Note that the Min-Total-Power relay selection can be seen as a special case of the adaptive selection rule with $C_i(t) = 1$ for all i and for all t.

Recognizing that some nodes may benefit more from the larger cooperative transmission opportunities than the others due to differences in the amount of data and to potentially unfair medium access protocol, we can generalize the rule even more to bring the balance (or "fairness") of the amount of payoffs that individual nodes collect.

Weighted Adaptive Relay Selection

A relay is selected for source i at time t such that

$$r_i(t) = \arg \min_{j \in \mathcal{R}_i(t), C_j(t)=1} \{w\left(u_j(0:t-1)\right)\left(p_{i,j}(t) + q_{i,j}(t)\right)\},$$

---

[4] We set $C_i(t) = 1$ if $u_i(0:t-1) = 0$ in order to enable the initial cooperative condition when all nodes' payoffs are zero. If $C_i(0) = 0$ for all i, no node would cooperate with other nodes.

where $w(u)$ is a non-increasing function of the payoff value u. Here, along with the power consumption factor $(p_{i,j}(t) + q_{i,j}(t))$, the weight function $w(u_j(0 : t-1))$ is introduced in the relay selection criteria, such that the nodes with larger payoffs (i.e., smaller weight) will have a higher chance to get selected as the relay for each packet transmission. More specifically, among relays that have the same total power consumption, preference will be given to the ones with higher cumulative payoff.

How much importance will be given to the weight term reflecting the payoff and how much to the power consumption term depends on how fast the function $w(u)$ decays as the payoff value u increases. For instance, one could use a power-law function $w(u) = u^{-k}$ with some positive constant k, and parameter k can be used to trade off fairness for energy consumption. In our simulation study, we find that $w(u) = u^{-6}$ strikes a good balance.

7.2.2 On Distributed Implementation of Relay Selection

We close this section by discussing how our relay selection rules can be realized in a distributed manner. Note that the relay selection rule requires the knowledge of (i) the estimates of the channel state information, and (ii) the current cumulative payoffs of the potential relays at the time of the packet transmission. In the following, we demonstrate how one can integrate our relay selection rule into a known medium access control (MAC) protocol—a similar idea can be used for other types of MAC protocols as well.

Specifically, we use a distributed protocol proposed in Adam et al. [41], which employs a four-way handshake of messages to control medium accesses for cooperative communication. In their protocol, when a source (S) attempts to transmit a message to a destination (D), a relay (R) is chosen by a random-access mechanism using the following message exchanges in sequence: (i) ready-to-send (RTS) sent by S, (ii) clear-to-send (CTS) sent by D, (iii) apply-for-relay (AFR) sent by R, and (iv) select-for-relay (SFR) sent by D. After the RTS and CTS messages, which serve the same role as in 802.11 MAC, an AFR message is broadcast by a relay (R) to notify other nodes of its intention to serve as the relay for S (SFR by D acknowledges AFR to avoid hidden relay problems). The way a particular relay is selected (and thus the selected relay sends AFR) is by having each potential relay back off for a random period of time.

Our relay selection rule can be readily implemented by innovatively using the above distributed protocol. We assume that all potential relays can hear the RTS and CTS messages.[5] First, the destination estimates the instantaneous channel quality of an S-D link from an RTS message and piggy-backs this information within the CTS message. Then all potential relays, upon hearing RTS and CTS messages, similarly estimate their respective S-R and R-D channel qualities and calculate the optimal power $p_{i,j}(t) + q_{i,j}(t)$ for the upcoming packet from S. Each relay then uses the calculated optimal power, along with its current payoff (should weighted adaptive relay selection be used), to set the backoff timer, proportional to $w(u_j(0 : t-1))(p_{i,j}(t) + q_{i,j}(t))$, and the node that sends the first AFR shall (and is implicitly chosen to) relay the packet from S.

7.2.3 Simulation Result

We evaluate the performance of our relay selection schemes via simulation, in which we place N (varied between 5 and 25) nodes at uniformly random locations in a 100 m $\times$ 100 m region (the edges of the region are wrapped (toroid) to eliminate edge effects). Throughout the simulation, we set the path-loss exponent $\alpha = 3$, the data rate $b = 1$ bps/Hz, and the targeted $\epsilon^{\text{out}} = 0.01$. A total of $200 \times N$ packets are transmitted, and at each time t, a packet is transmitted by a randomly selected source and a randomly selected destination. The initial payoff value of every node ($u_i(0)$) is set to 0. The simulation result is averaged over 100 times for each N.

Figure 7.4 shows the average energy consumption of each node, normalized by the minimum value in the data set (i.e., minimum total energy selection with 25 nodes) for different relay selection methods. Overall, our relay selection schemes perform far better than the direct transmission or that when a random relay is selected for each packet, and the adaptive relay-selection performs close to the minimum power selection, which is the optimal one in terms of average (or total) energy consumption (see Theorem 7.7). The weighted adaptive relay selection performs a bit worse (this is compensated by fairness results below). Furthermore, as the number of nodes increases, the average energy consumption of our relay selection schemes decreases; this is because it is easier to find a well-positioned relay and thus save more power in a dense network.

---

[5]This is a reasonable assumption since we consider cooperative relaying only when the S-R and R-D channels are good. Otherwise S will use direct transmission.

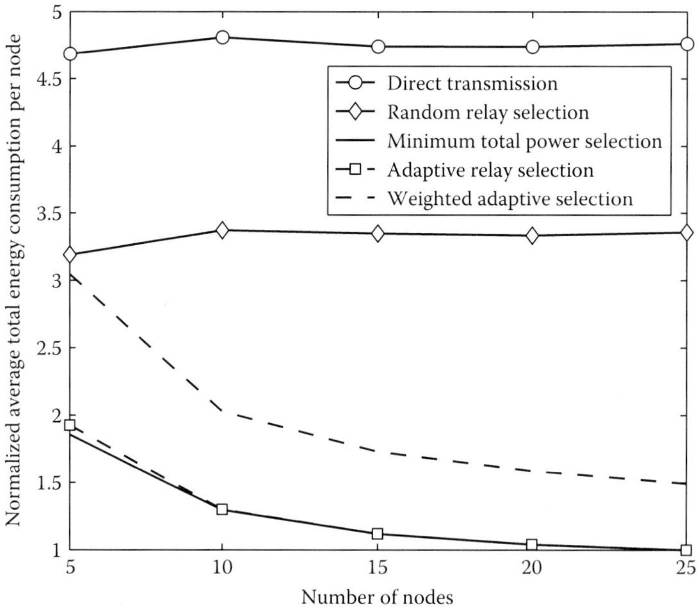

FIGURE 7.4 Average total energy consumption per node.

Figure 7.5 shows the fairness in terms of how much energy is saved for individual nodes using each relay selection method, where the y-axis represents Jain's fairness index of nodes' cumulative payoffs.[6] It is clear that the weighted adaptive relay-selection scheme achieves the best fairness compared with the other two schemes. As another example to highlight the fairness, we show in Figure 7.6 the energy consumptions of individual nodes at the end of simulation in a five-node network. It is clear that the weighted adaptive relay selection achieves the best fairness in this example—it is the only scheme that ensures that all nodes have positive payoff, whereas Min-Power-Selection results in negative payoff for some nodes.

We additionally conduct a set of simulations to measure the impact of each node's "willingness" to cooperate when its payoff is zero when our adaptive relay selection rule is used. To see this, we slightly changed the rule in Section 7.2.1 for a subset of nodes and divided the nodes into two groups: $\mathcal{U} = \{i \mid C_i(t) = 1 \text{ if } u_i(t) = 0\}$ ('Unselfish group'), and $\mathcal{S} = \{i \mid C_i(t) = 0 \text{ if } u_i(t) = 0\}$ ('Selfish group'); the rule remains the

[6]Jain's fairness index is defined by $(\sum u_i)^2/(N\sum u_i^2)$. The result ranges from $\frac{1}{N}$ (worst case) to 1 (best case). The larger the index is, the better fairness that we can achieve.

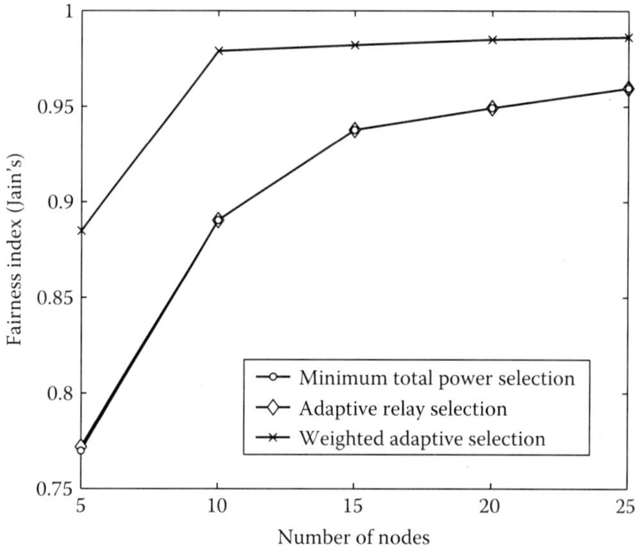

FIGURE 7.5 Jain's fairness on payoff function.

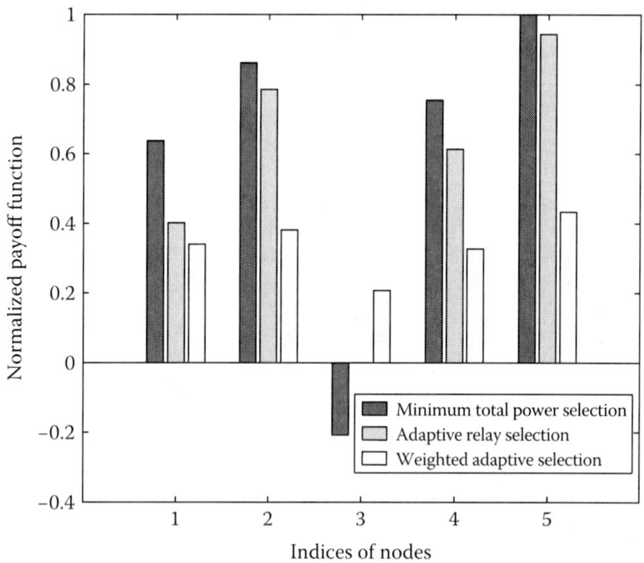

FIGURE 7.6 Normalized payoff function per node.

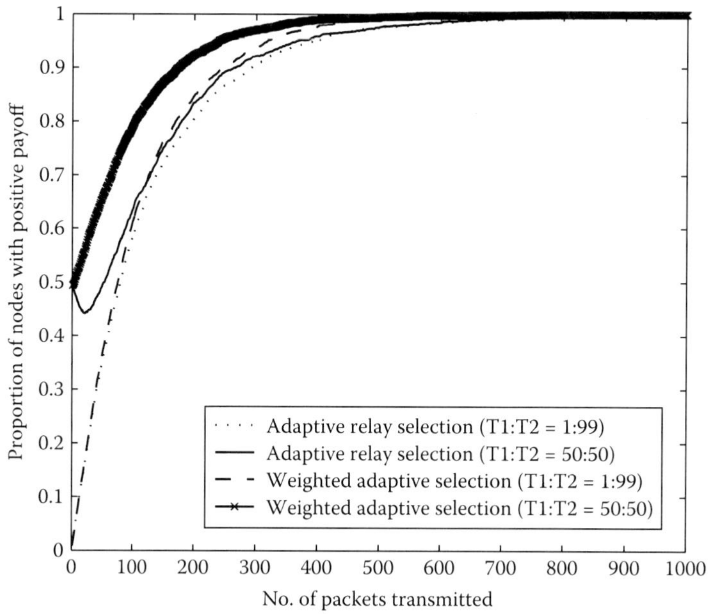

FIGURE 7.7 Proportion of nodes with positive payoff.

same as Section 7.2.1 for both groups when $u_i(t) \neq 0$, and the simulations are run using adaptive and weighted adaptive relay selection. We expect that the cooperative behaviors of the nodes tend to strengthen over time if more nodes are in the first group of "cooperative" nodes. In Figure 7.7, we show the proportion of the nodes with $C_i(t) = 1$ in the y-axis as the time progresses in the x-axis. Different curves represent different ratios of T_1:T_2, where $T_1 = |\mathcal{U}|$ and $T_2 = |\mathcal{S}|$ with 100 nodes in the network. The result is rather surprising: in all cases, the proportion of nodes in the cooperative states converges to 1, even when only one node cooperates initially to others out of 99. Also, convergence speed is faster with the weighted relay selection. What this result indicates is quite interesting: *the cooperative behavior of individuals can create a positive feedback loop between one another in cooperative communication*, and *the cooperation among the nodes can emerge even faster when combined with some policing mechanism for ensuring fair allocation of resources.*

III

Cooperative Communication in Multi-pair Multi-hop Scenario

CHAPTER 8

REACT: Residual Energy-Aware Cooperative Transmissions

Erwu Liu, Rui Wang, Chao Wang, Xinlin Huang, and Fuqiang Liu,
Tongji University, China

8.1 INTRODUCTION

Energy efficiency and spectrum efficiency are two critical issues in wireless communications. Previously, much effort has been on efficient use of the spectrum via various technologies such as compressive sensing and spectrum sensing [42]. On the other hand, a battery-operated wireless network must operate in an energy conservation manner so as to extend the lifetime of the network. Particularly in the development of the Internet of Things (IoT), energy efficiency is becoming equally or even more important in contrast to spectrum efficiency and prompting new waves of research activities. To achieve high energy efficiency, approaches have been proposed to deal with the problem from different angles across the layers from hardware up to application. These include efficient energy management techniques for peripheral devices on hand-held and embedded hardware platforms [43], energy-efficient medium access control design [44], energy-aware routing algorithm and topology generation [45, 46, 47], and energy-efficient algorithms for broadcast and multicast applications [48]. While each of these areas has received a lot of attention separately in recent years, a joint design of energy efficiency across the layers remains very limited due to the complexity of this problem. Our aim is to tackle the energy efficiency problem from cooperative transmission, relay selection, and scheduling perspectives in a battery-powered wireless network.

Cooperative diversity [49, 50, 51, 52, 53] is a new form of diversity through distributed transmission and processing with node collaboration. Transmit cooperation has nodes exchanging each other's messages, sharing their antennas, and creating multiple paths to transmit the information. Receive cooperation has nodes forwarding information about their observations for decoding. A system with both transmit and receive cooperation is similar to a multiple-input multiple-output (MIMO) system in a networked manner. Therefore it is sometimes called a distributed MIMO or network MIMO. It is known that transmit and receiver diversity can achieve higher capacity without sacrificing bandwidth or energy. In a network where each node is equipped with a single omnidirectional antenna, cooperative diversity can achieve similar gains from a MIMO system where each node is equipped with multiple antennas. However, there are several limitations that restrict the potential gains in a network MIMO compared to the conventional MIMO system, due to the absence of a direct high-capacity connection (e.g., usually a cable) among the antenna elements. For example, synchronization among the distributed antenna is much harder than a conventional MIMO transmitter. Additional resources such as bandwidth, power, and time are required to enable the cooperation. Antenna power allocation cannot be done as that in a conventional MIMO system and in other systems. In this research, we focus on the energy savings from cooperative diversity in an optimal way. Similar to Khandani et al. [54], we do not consider the effects of those limitations and assume that an appropriate architecture for achieving the required level of coordination among the cooperative nodes can take place [55].

Khandani et al. [54] propose a design of cooperative diversity to maximize energy savings. They formulate the problem as minimizing the overall energy consumption of the relay nodes between a transmission node and a receiving one. In this formulation, all relay nodes participate in cooperative transmission to gain energy savings. Further, relay nodes with very good channel conditions tend to participate more (e.g., transmit more often) than other nodes with poor channel conditions, since the good nodes are expected to use less transmission power than the poor nodes. This method achieves the overall savings of energy consumed by all the participating nodes. However, it may not be suitable for battery-powered nodes in an ad hoc or sensor network, where each node has its own limit of energy conservation. In fact, since nodes with good channel conditions transmit more often than others, these nodes will run out of their battery power more quickly. As a result, the number of cooperative relay nodes will decrease and cause an increase of the aggregate transmission power, which then again leads to quick power drain in the remaining good nodes.

Our goal is to address the similar energy-saving problem with consideration of each node's limited power conservation. We formulate the problem of energy savings as maximizing the lifetime of the network in terms of maximizing the overall number of packets transmitted by the source node to the destination, given a limited energy supply for the source node and each relay node along the optimal transmission path. To solve this problem, an intuitive solution is to select only a subset of relay nodes to participate in the cooperative transmission, thus avoiding the overuse of the good nodes in the set at each transmission stage. Unlike selective relaying in Bletsas et al. [51] where a relay with the highest capacity is selected, we construct a selection algorithm of the subset of nodes based on the residual energy of the nodes at each transmission stage. In our method, we trade a small portion of cooperative diversity gain with a much improved energy savings at each individual node, while still taking the wireless boardcast advantage (WBA) [48] and wireless cooperative advantage (WCA) [54]. Note that our method may not achieve the optimal energy savings in terms of total power consumption by all participating nodes as in the method of Khandani et al. [54]. However, our method will achieve a much improved lifetime in terms of preserving the residual energy of each node. Based on this consideration, we call it a residual energy-aware cooperative transmission (REACT) algorithm.

The rest of the chapter is structured as follows. In Section 8.2, we introduce basic terminology and concepts of cooperative transmission in relay-aided wireless networks; we then show that the traditional method is not suitable for battery-powered wireless networks. After that, REACT is proposed to avoid overly used relays while it still benefits from the traditional cooperative transmission techniques. Finally, in Section 8.3, we present simulation results to illustrate that the REACT algorithm is energy-efficient for both

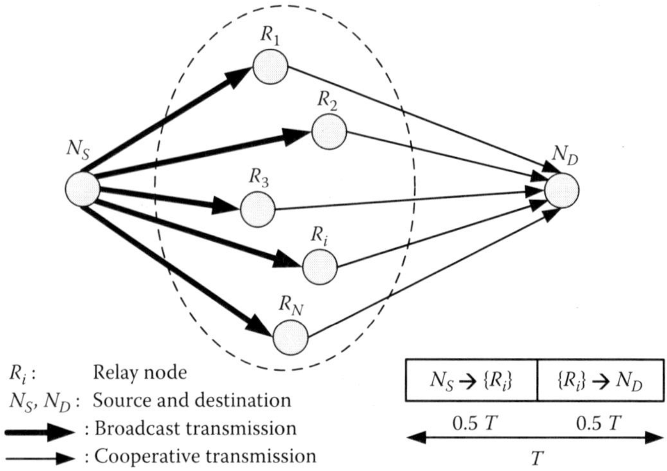

FIGURE 8.1 Cooperative transmission in wireless networks.

Rayleigh fading and non-fading environments, followed by the conclusion in Section 8.4.

8.2 SYSTEM MODEL

Refer to Figure 8.1 for a two-hop cooperative transmission. Without relays $R_1 \sim R_N$, nodes N_S and N_D are the one-hop neighbors in an ad hoc (or sensor) network. Consider the problem that node N_S wants to transmit packets to node N_D, with the help of N relays R_i $(i = 1, 2, \ldots, N)$. Since this two-hop cooperative transmission can be easily extended to the multi-hop case [54], here we focus on two-hop cooperative transmission. As shown in Figure 8.1, each transmission slot T is equally divided into two parts. At the first half slot, N_S broadcasts signals with power P_{BL} to relay nodes denoted as $R_1, R_2, \ldots, R_N$. At the second half slot, relay nodes that have successfully received and decoded the packets sent at the first half slot will cooperatively transmit to N_D. For a relay to successfully receive and decode signals over a broadcast link, we assume that the received signal-to-noise-ratio (SNR) must be at least $\text{SNR}_{\min}$. That is,

$$\frac{P_{BL} \times |h_{S,i}|^2}{N_0 \times W} \geq \text{SNR}_{\min}, \forall i, 1 \leq i \leq N, \tag{8.1}$$

where $h_{S,i}$ is the channel gain of the link $N_S \rightarrow R_i$, N_0 is the noise power density, and W is the bandwidth.

Similarly, for a destination node to successfully receive and decode signals over cooperative link, the received SNR must be at least $\text{SNR}_{\min}$. Khandani et al. [54] assumed in-phase receiving; that is, signal amplitudes are added at the destination node. We use the same assumption and have

$$\frac{\left(\sum_{i=1}^{N}(\sqrt{P_i} \times |h_{i,D}|)\right)^2}{N_0 \times W} \geq \text{SNR}_{\min}, \forall i, 1 \leq i \leq N, \tag{8.2}$$

where P_i is the transmit power of R_i and $h_{i,D}$ is the channel gain of link $R_i \rightarrow N_D$.

Given the above constraints, with Lagrangian multiplier techniques, one can prove that the aggregate transmit power of all relays is minimized if

$$P_i = \frac{|h_{i,D}|^2}{\text{SNR}_{\min} \times N_0 \times W} \bigg/ \left(\sum_{i=1}^{N} \frac{|h_{i,D}|^2}{\text{SNR}_{\min} \times N_0 \times W}\right)^2. \tag{8.3}$$

In Khandani et al. [54], the link cost of a wireless link is defined as the minimum power required to successfully transmit signals over the link. Using the same formulation, we have

$$LC_{S,i} = \frac{\text{SNR}_{\min} \times N_0 \times W}{|h_{S,i}|^2}, \forall i, 1 \leq i \leq N, \tag{8.4}$$

$$LC_{i,D} = \frac{\text{SNR}_{\min} \times N_0 \times W}{|h_{i,D}|^2}, \forall i, 1 \leq i \leq N, \tag{8.5}$$

$$LC_{BL} = \max_{1 \leq i \leq N} LC_{S,i}, \tag{8.6}$$

where $LC_{S,i}$, $LC_{i,D}$ are the link cost of links $N_S \rightarrow R_i$, $R_i \rightarrow N_D$ and LC_{BL} is the link cost of broadcast link $N_S \rightarrow \{R_i\}$.

Equation (8.3) can be rewritten as

$$P_i = \left(\frac{1}{LC_{i,D}}\right) \bigg/ \left(\sum_{i=1}^{N} \frac{1}{LC_{i,D}}\right)^2. \tag{8.7}$$

Denoted by LC_{CL}, the link cost of the cooperative link is

$$LC_{CL} = \sum_{i=1}^{N} P_i = 1 \Bigg/ \left(\sum_{i=1}^{N} \frac{1}{LC_{i,D}} \right). \tag{8.8}$$

We assume that the packet size is a bits. Denote η_S, η_i to be the required energy for N_S and R_i to transmit a packet, respectively. Obviously we have

$$\eta_S = LC_{BL} \times \frac{a}{W \times \log_2 (1 + \text{SNR}_{\min})}, \tag{8.9}$$

$$\eta_i = P_i \times \frac{a}{W \times \log_2 (1 + \text{SNR}_{\min})}. \tag{8.10}$$

In Khandani et al. [54], all relays participate in cooperative transmission and relays R_i transmit with power P_i given by (8.3). This scheme is not suitable for battery-powered ad hoc or sensor networks. According to (8.10), good-channel-condition relays (i.e., those with higher $|h_{i,D}|^2$) use more power in cooperative transmission, and these relays will run out of energy and die before bad-channel-condition ones with this scheme. When this happens, the number of cooperative relays will decrease and the aggregate power for cooperative transmission will increase, which makes the remaining relays die quickly. Eventually, N_S has to directly transmit to N_D as no relay is available and its lifetime could be greatly shortened in this scenario.

An intuitive solution to the above issue is to have those relays with relatively low residual energy not participate in cooperative transmission, while at the same time it should take the wireless broadcast advantage (WBA) [48] and wireless cooperative advantage (WCA) [54]. Based on this, we propose our residual energy-aware cooperative transmission (REACT) algorithm.

As shown in Figure 8.2, there are N relays $R_1, R_2, \ldots, R_N$, located between N_S and N_D. We use $\mathbf{RS} = \{R_i, 1 \leq i \leq M\}$ to denote the relay set of all active relays, where $M = |\mathbf{RS}|$ is the total number of relays in **RS**. Obviously we have $M = N$ initially. Unlike the work of Khandani et al. [54], which uses all active relays for cooperative transmission, the REACT algorithm chooses a subset of size k $(1 \leq k \leq M)$, $\mathbf{CS}_{j,k} \subseteq \mathbf{RS}$ to participate in one cooperative transmission at a time. Apparently, the REACT algorithm is simply the purely opportunistic transmission method (i.e., no cooperative transmission) for $k = 1$, and is simply the traditional cooperative transmission method (i.e., all relays participate in the cooperative transmission) for $k = M$.

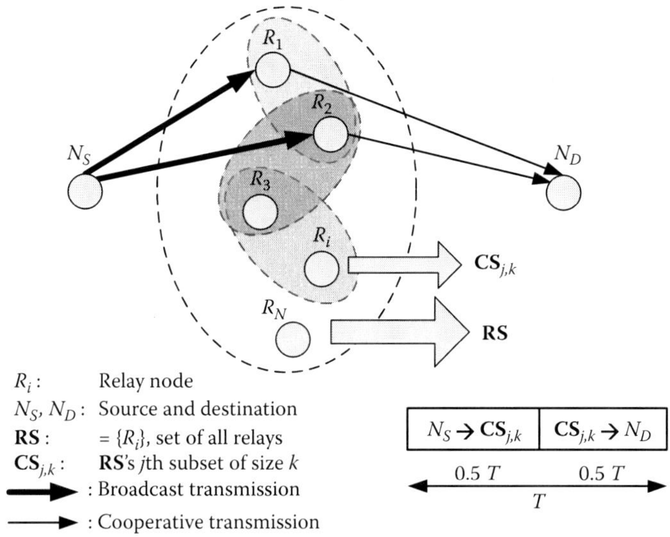

FIGURE 8.2 Residual Energy-Aware Cooperative Transmission (REACT).

When $\mathbf{CS}_{j,k}$ is chosen at slot t, Khandani's method [54] is used for the transmission; that is, N_S broadcasts to all relays in $\mathbf{CS}_{j,k}$ at the first half slot of t, and then all relays in $\mathbf{CS}_{j,k}$ cooperatively transmit to N_D at the second half slot of t. For the cooperative subset $\mathbf{CS}_{j,k}$, we use $LC_{BL,j}$ and $LC_{j,CL}$ to denote the link cost of the corresponding broadcast link and cooperative link, respectively, we use $\eta_{S,j}$ to denote the required energy for N_S to transmit a packet to $\mathbf{CS}_{j,k}$, and we use $P_{j,i}$ and $\eta_{j,i}$ to denote the required power and required energy, respectively, per packet of the ith relay in $\mathbf{CS}_{j,k}$. Similar to (8.6)–(8.10), we have

$$LC_{BL,j} = \max_{\forall R_i \in \mathbf{CS}_{j,k}} LC_{S,i}, \tag{8.11}$$

$$P_{j,i} = \left(\frac{1}{LC_{i,D}}\right) \Bigg/ \left(\sum_{\forall R_i \in \mathbf{CS}_{j,k}} \frac{1}{LC_{i,D}}\right)^2, \tag{8.12}$$

$$LC_{j,CL} = \sum_{\forall R_i \in \mathbf{CS}_{j,k}} P_{j,i} = 1 \Bigg/ \left(\sum_{\forall R_i \in \mathbf{CS}_{j,k}} \frac{1}{LC_{i,D}}\right), \tag{8.13}$$

$$\eta_{S,j} = LC_{BL,j} \times \frac{a}{W \times \log_2(1 + \text{SNR}_{\min})}, \tag{8.14}$$

$$\eta_{j,i} = P_{j,i} \times \frac{a}{W \times \log_2(1 + \text{SNR}_{\min})}. \tag{8.15}$$

We use $\varepsilon_S, \varepsilon_i$ to denote the residual energy of N_S and relay R_i ($1 \leq i \leq N$). We use V_j to denote the virtual number of packets that could be transmitted from N_S to $\mathbf{CS}_{j,k}$, and then to N_D using their current residual energy, that is,

$$V_j = \min\left(\frac{\varepsilon_S}{\eta_S}, \min_{\forall R_i \in \mathbf{CS}_{j,k}} \frac{\varepsilon_i}{\eta_i}\right), \tag{8.16}$$

where η_S, η_i are the required energy for N_S and R_i to transmit a packet under the current channel condition.

For an M-relay set **RS**, there are $\binom{M}{k}$ possible subsets of size k. For energy efficiency, we use the following criterion in choosing which subset for cooperative transmission:

$$i = \arg_{1 \leq j \leq \binom{M}{k}} \max V_j. \tag{8.17}$$

The above metric reveals that, given the residual energy, the cooperative set $\mathbf{CS}_{i,k}$ that maximizes the number of (virtually) transmittable packets will be selected. According to (8.16) and (8.17), when a relay has relatively low residual energy, the cooperative sets that contain this relay will have relatively low, (virtually) transmittable packet numbers, which would make them less likely to be chosen for cooperative transmission. By this, the REACT algorithm avoids overly used relays while it still benefits from the traditional cooperative transmission technique, thus improving efficiency. Obviously, REACT incorporates both opportunistic and cooperative transmission techniques, making it outperform the traditional cooperative transmission method.

Since $a/(W \times \log_2(1 + \text{SNR}_{\min}))$ is constant, the metric defined in (8.17) can be further expressed as

$$i = \arg_{1 \leq j \leq \binom{M}{k}} \max\left\{\min\left(\frac{\varepsilon_S}{LC_{BL,j}}, \min_{\forall R_i \in \mathbf{CS}_{j,k}} \frac{\varepsilon_i}{P_{j,i}}\right)\right\}. \tag{8.18}$$

Equation (8.18) is used in the REACT algorithm to determine which cooperative subset should be chosen for cooperative transmission. Algorithm 8.1 is the pseudo-code of the proposed REACT algorithm.

ALGORITHM 8.1 Residual Energy-Aware Cooperative Transmission (REACT) Algorithm

```
/* initial energy, energy threshold, packet size, k, bandwidth,
   minimum SNR, no. of packets per slot, channel gain at various
   slots                                                          */
Input: E_S, E_i, δ_S, δ_i, a, k, W, SNR_min, r, h_{S,D}, h_{S,i}, h_{i,D}, (1 ≤ i ≤ N)
/* m: total number of packets sent by the source N_S during its
   lifetime                                                       */
Output: m

RS = {R_i, 1 ≤ i ≤ N};
M = |RS|;
ε_S = E_S;
ε_i = E_i, ∀1 ≤ i ≤ N;
t = 1;                                            /* current slot */
m = 0;                               /* no. of packets transmitted */

for t ≥ 1 do
    LC_{S,i} = (SNR_min × N_0 × W)/|h_{S,i}[t]|^2, ∀1 ≤ i ≤ M;
    LC_{i,D} = (SNR_min × N_0 × W)/|h_{i,D}[t]|^2, ∀1 ≤ i ≤ M;
    if k > M then
        k = M;
    end
    if M ≥ 1 then
        /* relays available                                       */
        CS_{j,k} = RS's jth subset of size k, 1 ≤ j ≤ (M choose k);
        Calculate LC_{BL,j}, P_{j,i} according to (8.11) and (8.12);
        Choose the ith subset CS_{i,k} according to (8.18);
        /* N_S, CS_{i,k} consume energy                           */
        η_{S,i} = LC_{BL,i} × a/(W × log_2(1 + SNR_min));
        ε_S = ε_S − η_{S,i} × r;

        for ∀R_n ∈ CS_{i,k} do
            η_{i,n} = P_{i,n} × a/(W × log_2(1 + SNR_min));
            ε_n = ε_n − η_{i,n} × r;
            if ε_n < δ_n then
                /* no energy for R_n                              */
                remove R_n out of RS;
            end
        end
        m = m + r;                                /* add r packets */
    else
        /* no relay available, N_S directly transmits to N_D     */
        LC_{S,D} = (SNR_min × N_0 × W)/|h_{S,D}[t]|^2;
        η_{S,D} = LC_{S,D} × a/(W × log_2(1 + SNR_min));
        ε_S = ε_S − η_{S,D} × 2r;   /* tx 2r packets */ m = m + r;   /* add 2r
        packets */
    end
    if ε_S < δ_S then
        return(m);                           /* no energy for N_S */
    cnd
    M = |RS|;
    t = t + 1;                          /* proceeds to next slot */
end
```

In Algorithm 8.1, the initial energies of N_S, R_i are denoted as E_S, E_i. Considering that in practice, not all energy of a node/relay can be exclusively used for data transmission (e.g., some fraction of energy must be dedicated to ranging, carrier sensing, and/or signaling), we introduce energy threshold δ in the algorithm to indicate that when the residual energy of a node/relay falls below δ, the node/relay is considered dead and will not participate in the REACT algorithm. When a relay is dead, it will be removed from the relay set **RS**, and the algorithm will then recalculate all the $\binom{M}{k}$ possible cooperative subsets for the updated relay set **RS**. When all relays are dead, N_S will directly transmit to N_D. When the source node N_S is dead, the algorithm will terminate. We would like to point out that the input parameter r in the REACT algorithm is the number of packets that can be sent during a slot, that is, $r = T/(2a) \times W \times \log_2(1+\text{SNR}_{\text{min}})$ for relay-aided transmission ($N_S \to CS_{j,k} \to N_D$), and $r = T/a \times W \times \log_2(1+\text{SNR}_{\text{min}})$ for direct transmission ($N_S \to N_D$).

Next we evaluate the REACT algorithm.

8.3 SIMULATION RESULTS

In the simulations, the cooperative transmission method in Khandani et al. [54] is used as the baseline model for comparison. The channel gain in a wireless environment depends on the path-loss factor, the fast fading, and the slow fading (log-normal shadow fading). In most cases, these three phenomena are assumed to be independent. Fast fading is caused by multi-path propagation, while slow fading, or shadow fading, is caused by obstacles in the propagation path between two endpoints of a link. For the relay-aided transmission shown in Figure 8.1 or Figure 8.2, line-of-sight (LOS) communication between is typically assumed for both source-to-relay and relay-to-destination transmission. Hence, we do not need to consider the log-normal shadow fading here. In fact, the REACT algorithm presented here does not care whether there is fading or not.

We use the topology in Figure 8.3 for our simulation. There are eight relays located between N_S and N_D. Nodes and relays are placed in an area of 80×120 m^2. The simulation setup is as follows:

- Initial energy, $E_S = E_i = 10$ J.
- Minimum energy threshold, $\delta_S = \delta_i = 1$ J.

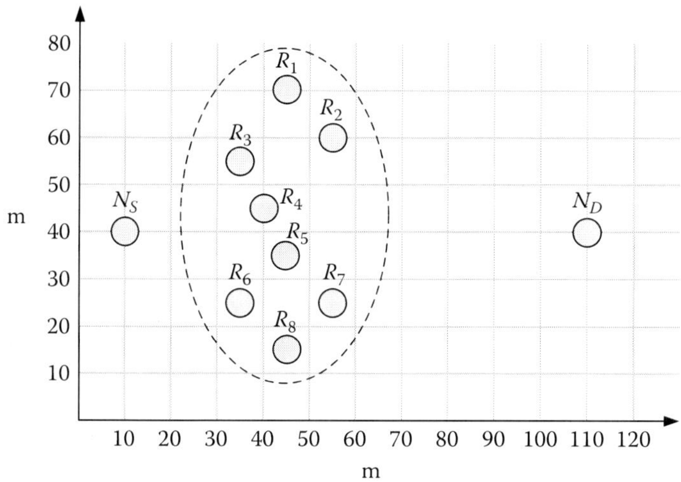

FIGURE 8.3 Relay-aided cooperative transmission in wireless networks.

- Bandwidth $W = 3.5$ MHz.
- $\text{SNR}_{\min} = 2.3$ dB. This corresponds to a rate of 5 Mbps.
- Packet size $a = 8000$ bits.
- Path-loss exponent $\alpha = 3.0$.
- Slot duration $T = 3.2$ ms.

We consider the following two simulation scenarios.

8.3.1 Path Loss Plus Rayleigh Fading

In this scenario, the path-loss factor and the fast Rayleigh fading contribute to the channel gain, that is,

$$h_{i,j} = \sqrt{PL_{d_0} \times \left(\frac{d_0}{d_{i,j}}\right)^{\alpha}} \times g_{i,j}, \tag{8.19}$$

where $g_{i,j}$ is the fast Rayleigh fading characterized by a zero-mean unit-variance complex Gaussian random variable and assumed to be independent for different nodes/relays, $PL_{d_0} \times (d_0/d_{i,j})^{\alpha}$ denotes the propagation loss of the transmission power, $d_{i,j}$ is the LOS distance from node/relay i to j, α is

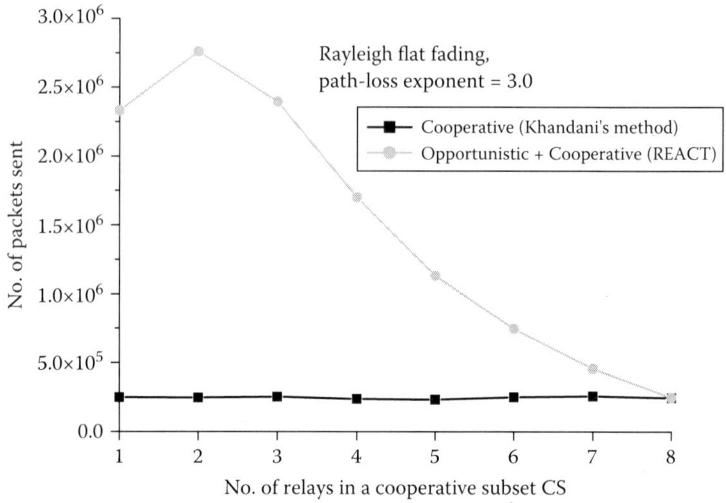

FIGURE 8.4 Total number of packets transmitted by N_S: REACT versus baseline method (fading case).

the path-loss exponent (typically 2–5), d_0 is the reference distance, and PL_{d_0} is the reference propagation loss at d_0.

Figure 8.4 illustrates the high efficiency of REACT. In Figure 8.4, we plot the number of packets transmitted by N_S during its lifetime for various sizes of cooperative sets. The REACT algorithm is the purely opportunistic transmission method (i.e., no cooperative transmission) for **CS** size $k=1$, and is the traditional cooperative transmission method (i.e., all relays participate in the cooperative transmission) for $k = 8$. The gray line (with circles) is the simulation result from the REACT algorithm while the black line (with squares) is the one from the baseline model. We can see that the improvement is significant. For example, the REACT algorithm produces about 11 times the number of packets transmitted by N_S during its lifetime when the size of **CS** is 2, and about 7 times when $k=4$ compared to the existing method.

It is known that the benefit of opportunistic resource allocation comes from channel fluctuation (e.g., fading). Similarly, the high efficiency of REACT is (partially) from fading as REACT opportunistically chooses a cooperative subset. We now evaluate the REACT algorithm when there is no fading.

8.3.2 Only Path Loss, No Fading

In this scenario, only path loss contributes to the channel gain, that is, $h_{i,j} = \sqrt{PL_{d_0} \times (d_0/d_{i,j})^{\alpha}}$. We use the same topology shown in Figure 8.3.

Similar to the fading case, we depict in Figure 8.5 the simulation results comparing our algorithm and the traditional one. Note that since all relays are involved in the transmission at each slot and there is no fading in this scenario, the simulation results for Khandani's method will remain the same for all simulation runs. From Figure 8.5 we can see that the performance improvement of the REACT algorithm is still remarkable in this case. In fact, the difference of the residual energy among nodes makes the selection of **CS** "opportunistic." In this sense, REACT artificially introduces "channel fluctuation" and thus improves performance even when there is no fading.

We also plot in Figure 8.6 the lifetime extension of REACT for various path-loss exponents, where the lifetime extension is defined as the ratio of the lifetime of N_S under the REACT algorithm to the one under the traditional method. We can see the REACT algorithm produces a lifetime extension of about 2.3–7.6 when the size of the cooperative subset is $\frac{1}{2}$ of the total number of relays. Figure 8.6 also shows that one or two relays are typically enough for REACT to perform well in fading scenarios.

In addition, we would like to point out that the benefit of cooperative transmission in REACT increases with k. On the other hand, the benefit of opportunistic transmission in REACT decreases with k when k is above some point. This trade-off between the cooperative transmission and opportunistic transmission is also seen in Figures 8.4, 8.5, and 8.6.

8.4 CONCLUSION

We have formulated the problem of maximizing the lifetime of a battery-operated node as maximizing the number of packets transmitted by the node. We have proposed a relay selection algorithm to choose a subset of relays in the relay set taking consideration of the residual power of each node. In our method, we avoid the selection of overly used nodes in each relay transmission stage, and thus the process is more energy efficient. Simulation results reveal much improvement in lifetime extensions with our REACT algorithm, compared with Khandani's method.

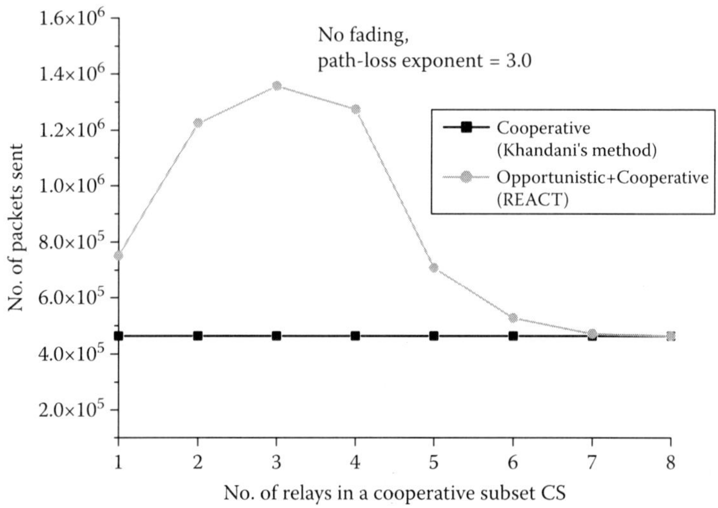

FIGURE 8.5 Total number of packets transmitted by N_S: REACT versus baseline method (no fading).

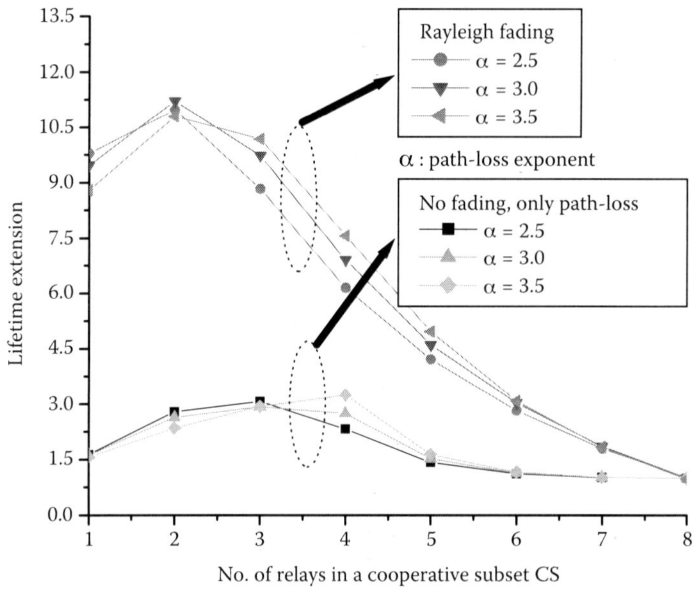

FIGURE 8.6 Lifetime extension of REACT.

CHAPTER 9

Joint Beamforming and Power Allocation

Chee Yen Leow, Universiti Teknologi Malaysia, Malaysia
Zhiguo Ding, Newcastle University
Kin K. Leung, Imperial College

9.1 INTRODUCTION

In real-life communication, data flows in both forward and backward link directions between source and destination. As an example, in a cellular network, downlink and uplink channels are used to support data flows in both link directions. This communication scenario is known as the information exchange channel. This chapter focuses on the study of the information exchange channel where both link directions are considered simultaneously in the system modeling.

In certain channel conditions, the direct link between the source and destination is unavailable to support two-way information exchange. This happens in situations such as when the source-to-destination channel is in a deep fade or undergoing severe shadowing, where the link quality is too weak to support any communication. In cellular systems, this also commonly occurs when the mobile user is located at cell edge, where the coverage of the base station is weak. In the absence of a direct communication link, the information exchange between a pair of users has to rely on the relay. Relay is a transceiver node placed in between the user pairs to help forward the data between users.

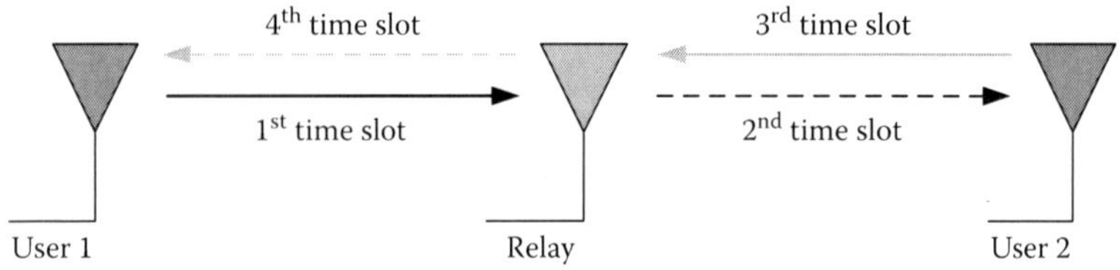

FIGURE 9.1 Information exchange using one-way relaying.

Using a conventional one-way relaying technique designed for unidirectional communication, the information exchange can only be completed in four channel uses due to half duplex constraint. Figure 9.1 shows the conventional one-way relaying scheme used for the information exchange. The information flows from user 1 to the relay, then from the relay to user 2 and vice versa, where a total of four time slots are used. This doubles the number of time slots used in direct point-to-point communication without a relay (when direct link between source and destination exists). In such an information exchange scenario, one-way relaying is spectrally inefficient because the achievable data rate is at most half of the data rate achievable by direct point-to-point communication.

Coincidentally, in the wired networking community, similar two-way information exchange scenarios have been addressed. An efficient technique known as network coding is first proposed in Ahlswede et al. [56]. The unique feature of network coding is that it allows intermediate nodes or relays to combine the information packets from multiple sources before forwarding to the destinations. This technique is shown to significantly enhance the overall network throughput [56].

Attracted by the benefit of network coding, two-way relaying has been proposed in wireless networks. Two-way relaying utilizes the broadcast nature of wireless transmission to enable data mixing between the user pair. Based on the original idea of network coding, two-way relaying is adapted in wireless networks to enhance the overall network throughput by reducing the channel resources used in the information exchange between users.

Two-way relaying schemes such as the DF-based scheme [57], analogue network coding [58], physical network coding [59], and so on, are able to complete the two-way information exchange in only two channel uses. Figure 9.2 explains the generic two-way relaying protocol in two time slots. In the first time slot, two users transmit simultaneously in the same channel to the relay. In the second time slot, the relay forwards the processed mixture

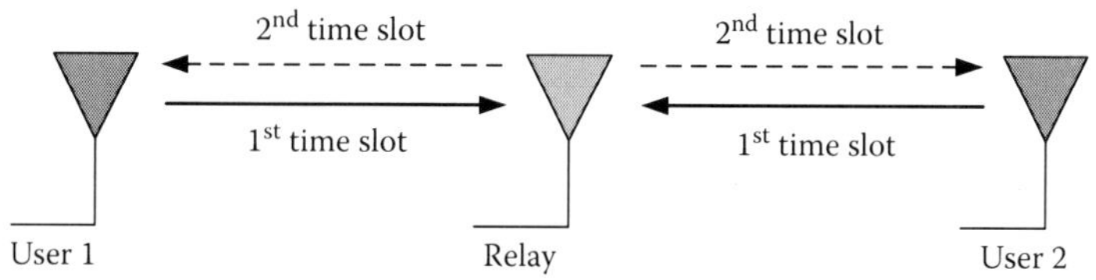

FIGURE 9.2 Information exchange using two-way relaying.

to the users and each user uses the knowledge of his previously transmitted message, known as self-interference, to decode the new message from his partner. Since the total channel use is halved compared to one-way relaying, the overall throughput can be doubled. This significant throughput enhancement motivates the application of two-way relaying in wireless networks.

For practical considerations, non-regenerative relaying, (i.e., amplify-and-forward (AF) relaying) is desirable when compared to regenerative relaying (i.e., decode-and-forward (DF)). This is due to the fact that non-regenerative relaying has lower complexity, has a lower processing delay, and incurs lower signal processing power compared to regenerative relaying.

To further enhance the system capacity and reliability, subsequent studies on non-regenerative two-way relaying extend to the multi-antenna scenario. Liang and Zhang [60] and Lee et al. [61] optimize the sum-rate using beamforming and power allocation techniques at the relay. The studies focus on the case where only the relay is equipped with multiple antennas. Li et al. [62] investigate the use of multiple relays and propose a relay beamforming technique to minimize the mean-squared-error (MSE). Leow et al. [63] study the scenario with multiple pairs of users and demonstrate that beamforming is able to remove co-channel interference and improve throughput and reliability.

The possibility of joint beamforming at the users and relay is explored in Lee et al. [64] and Xu and Hua [65]. It is found out in their work that the original sum-rate optimization problem is non-concave and therefore complicated to solve. An iterative searching algorithm is proposed in Lee et al. [64] to find locally optimal beamformers. The drawback is that the algorithm needs to be repeated extensively with different initial points to increase the probability of approaching the global optimal solution. Meanwhile, Xu and Hua [65] propose an alternate optimization technique to find locally optimal beamformers at the users when the beamformer at the relay is fixed, and vice versa until convergence. Similar to Lee et al. [64], the algorithm [65] also has to be repeated multiple times before it reaches the global optimal solution. This translates into high computation costs. In addition, Lee et al. [64] and Xu and Hua [65] only consider the case where each node is subject to fixed individual power constraints. The possible performance gain of implementing joint power allocation at all nodes subject to a total network power constraint remains unexplored.

This chapter considers a two-way relaying scenario that consists of a pair of multi-antenna users and a non-regenerative multi-antenna relay, all equipped with M antennas. A novel joint beamforming design based on subchannel alignment is proposed. The proposed beamforming design facilitates the investigation of the joint power allocation problem that maximizes the sum-rate, subject to a predefined total power constraint in the network. This problem is also known as the rate adaptive loading. Such network power allocation is critical in limiting the total interference incurred to a coverage area that is often regulated by the local authority. Numerical results show that the ergodic sum-rate of the proposed scheme significantly outperforms the baseline schemes.

The rest of the chapter is organized as follows. The system model and protocol description are presented in Section 9.2, while the beamforming design is discussed in Section 9.3. In Section 9.4, the joint power allocation problem is investigated. Section 9.5 covers the numerical simulation results. Section 9.6 concludes the chapter.

9.2 SYSTEM MODEL AND PROTOCOL DESCRIPTION

Consider a scenario where two users wish to exchange information with the help of a non-regenerative relay. The case where all nodes are equipped with M antennas is of interest. Figure 9.3 shows an example of the two-way relaying channel with $M = 2$. All channels are independently and identically distributed (i.i.d.) quasi-static Rayleigh distributed, and channel reciprocity is assumed. The receiver is corrupted by circularly symmetric additive white Gaussian noise. All nodes are subject to the half duplex constraint, which is realized through time division duplexing.

9.2.1 Initialization

Prior to the proposed information exchange protocol, the initialization phase takes place in order to enable all nodes to estimate the channels. The proposed

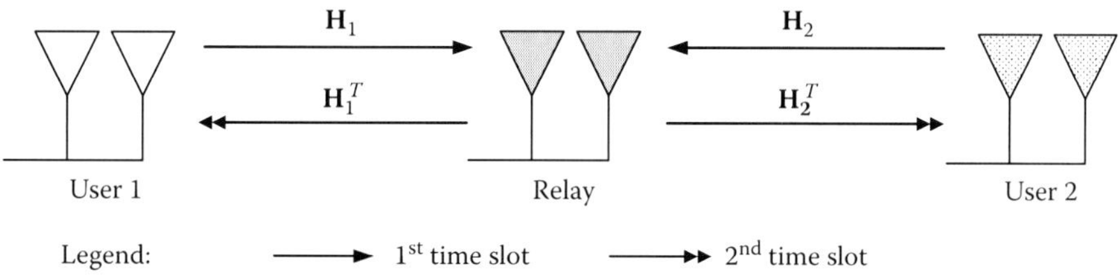

FIGURE 9.3 Example of two-way relaying scenario where each node is equipped with $M = 2$ antennas. The symbols above the arrows represent the channel matrices while the directions of the arrows indicate the directions of data flows.

two-way relaying protocol requires the relay to have full knowledge of the channel state information (CSI) of both relay-to-user channels, while each user knows his and his partner's user-to-relay channels. Having the CSI of the partnering channel (partner's user-to-relay channel) not only enables each user to decode the information from his partner but also facilitates each user to calculate suitable power allocation or power control factors to optimize the overall performance in the network.

Utilizing channel reciprocity, the CSI can be obtained using the open loop method [66], which can be accomplished within $3M + 1$ time slots. First, the relay uses M time slots to broadcast pilot sequences for both users to estimate their respective user-to-relay channel. Then, each user spends M time slots to broadcast pilot sequences so that the relay can estimate both relay-to-user channels. Finally, the relay consumes one time slot to broadcast the superposition of both users' CSI. Each user utilizes the knowledge of his local CSI to decode his partner's CSI. In comparison, conventional DF MIMO one-way relaying with receiver CSI requires $3M$ time slots in order to enable all nodes to estimate their receiver CSI. The proposed two-way relaying protocol only requires one extra time slot in acquiring the desired CSI.

9.2.2 Transmission Protocol

The proposed transmission protocol can be described in two time slots. Figure 9.3 summarizes the transmission flow of the proposed protocol. In the first time slot, both users transmit the linear precoded information vector to the relay; that is, user i transmits $\mathbf{F}_i\mathbf{x}_i$ where $\mathbf{F}_i \in \mathbb{C}^{M\times M}$ is the transmit beamforming matrix of user i and $\mathbf{x}_i \in \mathbb{C}^{M\times 1}$ is the information bearing vector of user i with normalized covariance (i.e., $E[\mathbf{x}_i\mathbf{x}_i^H] = \mathbf{I}_M$). The design of $\mathbf{F}_i$ will be discussed in the following section. The signal observed by the relay can be expressed as

$$\mathbf{r} = \mathbf{H}_1\mathbf{F}_1\mathbf{x}_1 + \mathbf{H}_2\mathbf{F}_2\mathbf{x}_2 + \mathbf{n}_r, \tag{9.1}$$

where $\mathbf{H}_i \in \mathbb{C}^{M\times M}$ is the channel from user i to the relay and $\mathbf{n}_r \in \mathbb{C}^{M\times 1}$ is the noise vector observed by the relay. In the second time slot, the relay broadcasts the linear precoded observation to both users, that is, $\mathbf{Wr}$ where $\mathbf{W} \in \mathbb{C}^{M\times M}$ is the joint receive and transmit beamforming matrix at the relay and $\mathbf{r} \in \mathbb{C}^{M\times 1}$ is the observation in the first time slot, expressed in (9.1). The relay beamforming matrix $\mathbf{W}$ will be discussed in the following section. The signal received by user i is

$$\mathbf{y}_i = \mathbf{H}_i^T\mathbf{W}\left(\mathbf{H}_1\mathbf{F}_1\mathbf{x}_1 + \mathbf{H}_2\mathbf{F}_2\mathbf{x}_2 + \mathbf{n}_r\right) + \mathbf{n}_i, \tag{9.2}$$

where $\mathbf{n}_i \in \mathbb{C}^{M\times 1}$ is the noise vector observed by user i. Each user performs linear post-processing to the received mixture broadcast by the relay, that is, user i calculates $\mathbf{G}_i\mathbf{y}_i$ where $\mathbf{G}_i \in \mathbb{C}^{M\times M}$ is the receive beamforming matrix for user i. Self-interference of user i contains the information transmitted by user i himself in the previous time slot, that is, $\mathbf{G}_1\mathbf{H}_1^T\mathbf{W}\mathbf{H}_1\mathbf{F}_1\mathbf{x}_1$ is the self-interference of user 1. Using the principle of analogue network coding [58], the self-interference is subtracted from the mixture and the desired information vector can be decoded. Writing $\mathbf{A}_i = \mathbf{G}_i\mathbf{H}_i^T\mathbf{W}\mathbf{H}_{\bar{i}}\mathbf{F}_{\bar{i}}$ and $\mathbf{B}_i = \mathbf{G}_i\mathbf{H}_i^T\mathbf{W}$, and assuming Gaussian channel coding, the mutual information of user i can be expressed as

$$R_i = \frac{1}{2}\log_2\left(\det\left(\mathbf{I}_M + \mathbf{A}_i\mathbf{A}_i^H\left(\sigma_r^2\mathbf{B}_i\mathbf{B}_i^H + \sigma_i^2\mathbf{G}_i\mathbf{G}_i{}^H\right)^{-1}\right)\right), \tag{9.3}$$

where σ_r^2 is the receiver noise power at the relay, σ_i^2 is the receiver noise power at user i, and the subscript $\bar{i}$ is used to denote the complement of i, that is, when $i = 1$, $\bar{i} = 2$. The pre-log factor reflects the two time slots used to complete the information exchange.

9.3 BEAMFORMING DESIGN

In this section, the proposed low-complexity design of transmit beamformer at the users $\mathbf{F}_i$, $\forall i = \{1, 2\}$, joint receive and transmit beamformer at the relay $\mathbf{W}$, and receive beamformer at the users $\mathbf{G}_i$, $\forall i = \{1, 2\}$ are described. The objective of the beamforming design is to decompose the channels into M parallel subchannels, which not only facilitates substream power allocation and power control but also enables the use of simple SISO decoders at the users.

9.3.1 Design of $\mathbf{F}_i$

Recall that in a conventional point-to-point MIMO system, the optimal transmit and receive beamformers are designed by means of singular value decomposition (SVD), such that the channel matrix is decomposed into parallel subchannels (or eigenmodes) to enable optimal power sharing among subchannels [66]. However, this cannot be directly implemented in the two-way relaying scenario considered here, as mentioned in Lee et al. [64]. This is due to the fact that the relay (which acts as a MIMO receiver) is not able to simultaneously separate the subchannels from user 1 and user 2, which have different channel directions.

To address the above issue, subchannel alignment is proposed in the design of $\mathbf{F}_i$ to ensure that the kth subchannel of user 1 and the kth subchannel of user 2 occupy the same signal subspace. Specifically, the following structure for the transmit beamformer of user i is proposed:

$$\mathbf{F}_i = \tilde{\mathbf{F}}_i \mathbf{V}_i \mathbf{\Sigma}_i, \tag{9.4}$$

where alignment matrix $\tilde{\mathbf{F}}_i \in \mathbb{C}^{M \times M}$ is obtained from subchannel alignment,[1]

$$\mathbf{H}_1 \tilde{\mathbf{F}}_1 = \mathbf{H}_2 \tilde{\mathbf{F}}_2. \tag{9.5}$$

The subchannel alignment problem can be solved as follows:

$$\begin{bmatrix} \tilde{\mathbf{F}}_1 \\ -\tilde{\mathbf{F}}_2 \end{bmatrix} = \text{null}\left(\begin{bmatrix} \mathbf{H}_1 & \mathbf{H}_2 \end{bmatrix}\right), \tag{9.6}$$

[1]Note that $\tilde{\mathbf{F}}_i$ is not unique. However, multiplication of $\tilde{\mathbf{F}}_i$ with a unitary matrix (rotation matrix) does not change the singular values of the effective channels, that is, $\mathbf{\Lambda}_i$ remains the same.

where the computation of the null-space vectors can be found in Strang [67]. Matrix $\mathbf{V}_i \in \mathbb{C}^{M\times M}$ in (9.4) is the right singular matrix obtained from the SVD of $\mathbf{H}_i\tilde{\mathbf{F}}_i$; that is, $\mathbf{H}_i\tilde{\mathbf{F}}_i = \mathbf{U}_i\mathbf{\Lambda}_i\mathbf{V}_i^H$, where $\mathbf{U}_i \in \mathbb{C}^{M\times M}$ is the left singular matrix and $\mathbf{\Lambda}_i \in \mathbb{R}^{M\times M}$ is the diagonal matrix of singular values. The diagonal matrix $\mathbf{\Sigma}_i \in \mathbb{R}^{M\times M}$ in (9.4) is the transmit power allocation matrix of user i. The transmit power consumption of user i is $\|\mathbf{F}_i\|_F^2$. Due to subchannel alignment in (9.5), $\mathbf{U}_1\mathbf{\Lambda}_1\mathbf{V}_1^H = \mathbf{U}_2\mathbf{\Lambda}_2\mathbf{V}_2^H$. The subscripts of $\mathbf{U}_i\mathbf{\Lambda}_i\mathbf{V}_i^H$ are omitted for simplicity of notation. The received signal at the relay in (9.1) reduces to

$$\mathbf{r} = \mathbf{U\Lambda}\left(\mathbf{\Sigma}_1\mathbf{x}_1 + \mathbf{\Sigma}_2\mathbf{x}_2\right) + \mathbf{n}_r. \tag{9.7}$$

9.3.2 Design of W

The design of joint receive and transmit beamformer $\mathbf{W}$ ensures that the received signal in (9.7) can be decomposed into parallel streams. To achieve the objective of subchannel decomposition, the following structure is proposed:

$$\mathbf{W} = \mathbf{U}^*\mathbf{\Sigma}_r\mathbf{U}^H, \tag{9.8}$$

where $\mathbf{U}$ is the left singular matrix of $\mathbf{H}_i\tilde{\mathbf{F}}_i$, and diagonal matrix $\mathbf{\Sigma}_r \in \mathbb{R}^{M\times M}$ is the transmit power allocation matrix at the relay. Notice that the use of subchannel alignment discussed in the previous subsection enables the relay to decompose the channels of user 1 and user 2 simultaneously. The total transmit power consumption at the relay is $\|\mathbf{WH}_1\mathbf{F}_1\|_F^2 + \|\mathbf{WH}_2\mathbf{F}_2\|_F^2 + \sigma_r^2\,\|\mathbf{W}\|_F^2$, where σ_r^2 is the noise power in watts at the relay.

9.3.3 Design of $\mathbf{G}_i$

The design of the receive beamformer $\mathbf{G}_i$ is to ensure that the received signal in (9.2) can be decomposed into parallel streams. Specifically, the following structure is proposed:

$$\mathbf{G}_i = \mathbf{V}^T\tilde{\mathbf{F}}_i^T, \tag{9.9}$$

which is the transposition of the transmit beamforming matrix $\mathbf{F}_i$ but without the power allocation matrix. The signal received by user i in (9.2) after the receive beamforming in (9.9) is applied simplifies to

$$\mathbf{G}_i\mathbf{y}_i = \mathbf{\Lambda}^2\mathbf{\Sigma}_r\left(\mathbf{\Sigma}_1\mathbf{x}_1 + \mathbf{\Sigma}_2\mathbf{x}_2\right) + \tilde{\mathbf{n}}_i, \tag{9.10}$$

where $\tilde{\mathbf{n}}_i = \mathbf{\Lambda}\mathbf{\Sigma}_r\mathbf{U}^H\mathbf{n}_r + \mathbf{V}^T\tilde{\mathbf{F}}_i^T\mathbf{n}_i$ is the effective noise observed by user i. As described in Section 9.2, each user is able to decode the desired information by subtracting the self-interference from the mixture. For instance, user 1 is able to decode $\mathbf{x}_2$ by subtracting the self-interference, $\mathbf{\Lambda}^2\mathbf{\Sigma}_r\mathbf{\Sigma}_1\mathbf{x}_1$, from the received mixture.

9.4 JOINT POWER ALLOCATION

This section investigates the joint power allocation problem of the proposed beamforming scheme. First, the SNR of each subchannel is derived. Second, the joint power allocation problem is formulated using the sum-rate criterion, and the convexity of the optimization problem is verified. Since the objective function is non-concave, an upper bound is derived to approximate the original objective function. The last subsections discuss the proposed power allocation strategies, baseline schemes, and a comparable scheme.

9.4.1 Subchannel SNR Derivation

From (9.10), it can be observed that the channel matrices are decomposed into M parallel subchannels. In this subsection, the SNR of each subchannel is derived. Denote the transmit power allocation matrix of user 1, $\mathbf{\Sigma}_1 = \text{diag}(\ \sqrt{a_1}\ \ \dots\ \ \sqrt{a_M}\ )$, the transmit power allocation matrix of user 2, $\mathbf{\Sigma}_2 = \text{diag}(\ \sqrt{b_1}\ \ \dots\ \ \sqrt{b_M}\ )$, the transmit power allocation matrix of the relay, $\mathbf{\Sigma}_r = \text{diag}(\ \sqrt{c_1}\ \ \dots\ \ \sqrt{c_M}\ )$, the diagonal matrix of singular values, $\mathbf{\Lambda} = \text{diag}(\ \sqrt{\lambda_1}\ \ \dots\ \ \sqrt{\lambda_M}\ )$, and $\hat{\mathbf{F}}_i = \tilde{\mathbf{F}}_i\mathbf{V}_i$. The variables a_k, b_k, and c_k represent the kth substream power allocation factors for user 1, user 2, and the relay, respectively. Assuming a SISO decoder is used to decode each parallel stream, the SNR of the kth subchannel of user 1 can be expressed as follows:

$$\gamma_{1,k} = \frac{\lambda_k^2 b_k c_k}{\sigma_r^2\lambda_k c_k + \sigma_1^2\sum_{j=1}^{M}|\hat{\mathbf{F}}_1(j,k)|^2}, \tag{9.11}$$

where σ_1^2 is the noise power at the user 1 receiver. Similarly, the SNR of the kth subchannel of user 2 can be written as

$$\gamma_{2,k} = \frac{\lambda_k^2 a_k c_k}{\sigma_r^2\lambda_k c_k + \sigma_2^2\sum_{j=1}^{M}|\hat{\mathbf{F}}_2(j,k)|^2}, \tag{9.12}$$

where σ_2^2 is the noise power at the user 2 receiver. Assuming Gaussian channel coding, the instantaneous data rate (or mutual information) of user i can be expressed as

$$R_i = \frac{1}{2} \sum_{k=1}^{M} \log_2 \left(1 + \gamma_{i,k}\right), \tag{9.13}$$

where the pre-log factor reflects the two time slots used to complete the information exchange. In a realistic wireless system using a practical channel coding scheme, the instantaneous data rate of user i can be modified as

$$R_i' = \frac{1}{2} \sum_{k=1}^{M} \log_2 \left(1 + \frac{\gamma_{i,k}}{\Gamma}\right), \tag{9.14}$$

which includes the SNR gap Γ to account for the target error probability and the specific channel coding scheme [68]. The SNR gap has a typical value of $\Gamma \geq 1$ where the special case $\Gamma = 1$ corresponds to the upper bound in (9.13) where Gaussian coding is used. Since Γ is independent of the channel, the formulation in this chapter assumes $\Gamma = 1$ without loss of generality.

9.4.2 Sum-Rate Optimization

Sum-rate criterion (i.e., $R_1 + R_2$) is the optimization utility used in this chapter. All transmissions in the network are subject to a total network power constraint P in watts, which can be written as

$$\sum_{i=1}^{2} \left(\|\mathbf{F}_i\|_F^2 + \|\mathbf{W}\mathbf{H}_i\mathbf{F}_i\|_F^2\right) + \sigma_r^2 \|\mathbf{W}\|_F^2 \leq P. \tag{9.15}$$

The joint power constraint is the summation of the transmit power consumption at user 1, user 2, and the relay. The joint power constraint expression can be simplified. Specifically, the transmit power consumption at the relay can be simplified as $\|\mathbf{W}\mathbf{H}_1\mathbf{F}_1\|_F^2 + \|\mathbf{W}\mathbf{H}_2\mathbf{F}_2\|_F^2 + \sigma_r^2 \|\mathbf{W}\|_F^2 = \sum_{k=1}^{M} \left(\lambda_k a_k c_k + \lambda_k b_k c_k + \sigma_r^2 c_k \sum_{j=1}^{M} |\mathbf{U}(j,k)|^2\right)$. Note that $\sum_{j=1}^{M} |\mathbf{U}(j,k)|^2 = 1$. Similarly, one can simplify $\|\mathbf{F}_1\|_F^2 = \sum_{k=1}^{M} \sum_{j=1}^{M} a_k |\hat{\mathbf{F}}_1(j,k)|^2$ and $\|\mathbf{F}_2\|_F^2 = \sum_{k=1}^{M} \sum_{j=1}^{M} b_k |\hat{\mathbf{F}}_2(j,k)|^2$. To further simplify the expression, one can represent $\tilde{a}_k = a_k \sum_{j=1}^{M} |\hat{\mathbf{F}}_1(j,k)|^2$, $\tilde{b}_k = b_k \sum_{j=1}^{M} |\hat{\mathbf{F}}_2(j,k)|^2$, and $\tilde{c}_k = c_k \left(\lambda_k a_k + \lambda_k b_k + \sigma_r^2\right)$ as the effective kth substream power allocation factors for user 1, user 2, and the relay, respectively.

The joint power allocation problem using sum-rate criterion[2] can be formulated as

$$\underset{\tilde{a}_1,\ldots,\tilde{a}_M,\tilde{b}_1,\ldots,\tilde{b}_M,\tilde{c}_1,\ldots,\tilde{c}_M}{\text{maximize}} \quad \frac{1}{2}\sum_{k=1}^{M}\log_2\left(1+\frac{t_{1,k}\tilde{b}_k\tilde{c}_k}{t_{2,k}\tilde{a}_k+t_{3,k}\tilde{b}_k+t_{4,k}\tilde{c}_k+t_{5,k}}\right)$$
$$+\frac{1}{2}\sum_{k=1}^{M}\log_2\left(1+\frac{u_{1,k}\tilde{a}_k\tilde{c}_k}{u_{2,k}\tilde{a}_k+u_{3,k}\tilde{b}_k+u_{4,k}\tilde{c}_k+u_{5,k}}\right), \tag{9.16}$$

$$\text{subject to} \quad \sum_{k=1}^{M}\left(\tilde{a}_k+\tilde{b}_k+\tilde{c}_k\right)\leq P, \tilde{a}_k\geq 0, \tilde{b}_k\geq 0, \tilde{c}_k\geq 0,$$
$$\forall_k=\{1,\ldots,M\}, \tag{9.17}$$

where the constants

$$t_{1,k}=\lambda_k^2, \qquad t_{2,k}=\sigma_1^2\beta_k\lambda_k, \qquad t_{3,k}=\sigma_1^2\alpha_k\lambda_k,$$
$$t_{4,k}=\sigma_r^2\beta_k\lambda_k, \qquad t_{5,k}=\sigma_1^2\sigma_r^2\alpha_k\beta_k, \qquad u_{1,k}=t_{1,k}=\lambda_k^2,$$
$$u_{2,k}=\sigma_2^2\beta_k\lambda_k, \qquad u_{3,k}=\sigma_2^2\alpha_k\lambda_k, \qquad u_{4,k}=\sigma_r^2\alpha_k\lambda_k,$$
$$u_{5,k}=\sigma_2^2\sigma_r^2\alpha_k\beta_k, \qquad \alpha_k=\sum_{j=1}^{M}|\hat{\mathbf{F}}_1(j,k)|^2, \qquad \beta_k=\sum_{j=1}^{M}|\hat{\mathbf{F}}_2(j,k)|^2.$$

The optimization problem can be solved using convex optimization techniques [69] if the constraints are convex and the objective is concave. The inequality of the power constraint in (9.17) is affine, hence convex (and concave), with respect to all input parameters $\tilde{a}_1,\ldots,\tilde{a}_M,\tilde{b}_1,\ldots,\tilde{b}_M,\tilde{c}_1,\ldots,\tilde{c}_M$. However, it can be shown that the objective function in (9.16) is non-concave with respect to all input parameters.

[2]The sum-rate optimization in (9.16) with constraints in (9.17) can be easily modified to include the special case where each node has fixed power constraint. Specifically, the joint power constraint in (9.17) is separated into three individual constraints, i.e., $\sum_{k=1}^{M}\tilde{a}_k=P_1$, $\sum_{k=1}^{M}\tilde{b}_k=P_2$, and $\sum_{k=1}^{M}\tilde{c}_k=P_r$, where P_1, P_2, and P_r are the individual power constraint at user 1, user 2, and the relay, respectively.

It is shown that the objective function in (9.16) is non-concave with respect to the parameters $\tilde{a}_1, \ldots, \tilde{a}_M, \tilde{b}_1, \ldots, \tilde{b}_M, \tilde{c}_1, \ldots, \tilde{c}_M$. In order to ease the difficulty in solving the power allocation problem, a concave upper bound of the original objective function that can be solved efficiently using convex optimization techniques is derived. The following result summarizes the concavity of the derived upper bound.

The upper bound of the objective function is jointly concave with respect to input parameters $\tilde{a}_1, \ldots, \tilde{a}_M, \tilde{b}_1, \ldots, \tilde{b}_M, \tilde{c}_1, \ldots, \tilde{c}_M$,

$$
\begin{aligned}
f_{\text{upper}} = & \frac{1}{2} \sum_{k=1}^{M} \log_2 \left(1 + \frac{\left(\tilde{a}_k + \tilde{b}_k\right) \tilde{c}_k}{t_{a,k} \left(\tilde{a}_k + \tilde{b}_k\right) + t_{b,k} \tilde{c}_k} \right) \\
& + \frac{1}{2} \sum_{k=1}^{M} \log_2 \left(1 + \frac{\left(\tilde{a}_k + \tilde{b}_k\right) \tilde{c}_k}{u_{a,k} \left(\tilde{a}_k + \tilde{b}_k\right) + u_{b,k} \tilde{c}_k} \right), \qquad (9.18)
\end{aligned}
$$

where the constants

$$
t_{a,k} = \frac{\min(t_{2,k}, t_{3,k})}{t_{1,k}}, \qquad t_{b,k} = \frac{t_{4,k}}{t_{1,k}},
$$

$$
u_{a,k} = \frac{\min(u_{2,k}, u_{3,k})}{u_{1,k}}, \qquad u_{b,k} = \frac{u_{4,k}}{u_{1,k}}.
$$

Remark The power allocation factors, $\tilde{a}_1, \ldots, \tilde{a}_M, \tilde{b}_1, \ldots, \tilde{b}_M, \tilde{c}_1, \ldots, \tilde{c}_M$, obtained by solving the concave upper bound in (9.18) are suboptimal solutions to the original problem in (9.16). The approximation in (9.18) enables the transmission power to be allocated dynamically between users and relay, while the users share identical power allocation factors. Since the positive sum of the power allocation factors of both users (i.e., $\tilde{a}_k + \tilde{b}_k$), can be represented as a single power allocation factor, the result is a combination of dynamic power sharing between users and relay, coupled with equal power sharing between users.

9.4.3 Proposed Power Allocation Strategies

In this subsection, two joint power allocation (JPA) strategies are proposed.

Proposed JPA I

The proposed JPA I computes the power allocation factors by solving the sum-rate optimization problem in (9.16) and (9.17). As discussed in the previous subsection, the objective function in (9.16) is non-concave. In this case, the locally optimal solution does not necessarily correspond to the globally optimal solution. The globally optimal solution can be found with a certain probability by means of randomization-based global optimization [70]. For each channel realization, multiple random starting vectors are generated and the local optimal solution for each starting vector is computed using convex optimization techniques, that is, the interior-point method [69]. The globally optimal solution for each channel realization is the maximum of all local optimal solutions. Since this method requires the use of multiple random starting vectors, a centralized node (i.e., the relay) will compute the power allocation factors and distribute them to other nodes.

Proposed JPA II

The proposed JPA II computes the power allocation factors by solving the concave upper bound in (9.18). The power allocation factors can be calculated efficiently using the convex optimization techniques, that is, the interior-point method [69]. Since the upper bound objective function in (9.18) is concave, the local optimal solution obtained using convex optimization corresponds to the global optimal solution. The computed power allocation factors corresponding to the globally optimal solution of (9.18) are then substituted back into the original objective function in (9.16) to obtain the achievable sum-rate. With the CSI knowledge of the channels, each node is able to compute the power allocation factors locally, without any cooperation between nodes. Specifically, perfect knowledge of both $\mathbf{H}_1$ and $\mathbf{H}_2$ enables every node to have common knowledge of all the constants in (9.18). This allows all nodes to compute the solution to the same optimization problem locally. The CSI training scheme to enable all nodes to have the CSI of both $\mathbf{H}_1$ and $\mathbf{H}_2$ is discussed in Subsection 9.2.1.

9.4.4 Baseline Schemes and Comparable Scheme

In this subsection, two baseline schemes, pure AF MIMO two-way relaying and DF MIMO one-way relaying schemes, are presented. The best comparable scheme [65] is also discussed.

Baseline Scheme: Pure AF

In the pure AF two-way relaying scheme, the relay simply forwards the power normalized observation to the users, without any beamforming or power allocation. Assuming optimal MIMO decoders and equal power allocation, sum-rate can be computed using (9.3) with

$$\mathbf{F}_i = \sqrt{\frac{P_i}{M}}\mathbf{I}, \qquad \mathbf{G}_i = \mathbf{I},$$

$$\mathbf{W} = \sqrt{\frac{P_r}{\|\mathbf{H}_1\mathbf{F}_1\|_F^2 + \|\mathbf{H}_2\mathbf{F}_2\|_F^2 + \sigma_r^2 M}}\mathbf{I},$$

where P_i is the power constraint at user i and P_r is the power constraint at the relay. It is assumed that $P_1 = P_2 = P_r = \frac{P}{3}$ so that the total network power constraint P is satisfied. This scheme serves as a baseline to study the contribution of power allocation to the sum-rate of the two-way relaying channel.

Baseline Scheme: MIMO One-Way

Another baseline scheme used for comparison is the DF MIMO one-way relaying. Recall that in one-way relaying, four orthogonal channel uses are consumed to complete the information exchange between user pairs. In the first time slot, user 1 transmits information to the relay. After decoding the received information, the relay forwards the observation to user 2 in the second time slot. Following a similar fashion, user 2 transmits information to user 1 with the help of relay using another two time slots.

Assuming the perfect transmitter and receiver CSI are available, the channel matrix from user i to relay, $\mathbf{H}_i$, can be decomposed into M parallel streams using the SVD, that is, $\mathbf{H}_i = \mathbf{U}_i\mathbf{\Lambda}_i\mathbf{V}_i^H$ where $\mathbf{\Lambda}_i = \text{diag}(\sqrt{\lambda_{i,1}}, \ldots, \sqrt{\lambda_{i,M}})$. Note that the $\mathbf{U}_i\mathbf{\Lambda}_i\mathbf{V}_i^H$ and $\lambda_{i,k}$ defined here are different from those in Subsection 9.3.1 and 9.4.1. Denote a_k and b_k as the kth substream power allocation factor of user 1 and user 2, respectively. Variable c_k is defined as the kth substream power allocation factor of the relay for transmission to user 1, while variable d_k is the kth substream power allocation factor of the relay for transmission to user 2. Represent $R_{i\rightarrow r}$ as the data rate from user i to relay and $R_{r\rightarrow i}$ as the data rate from the relay to user i.

The data rate at user 1 can be expressed as

$$\mathcal{I}_1 = \min(R_{2\to r}, R_{r\to 1}) \tag{9.19}$$

$$= \min\left(\frac{1}{4}\sum_{k=1}^{M}\log_2\left(1+\frac{b_k\lambda_{2,k}}{\sigma_r^2}\right), \frac{1}{4}\sum_{k=1}^{M}\log_2\left(1+\frac{c_k\lambda_{1,k}}{\sigma_1^2}\right)\right), \tag{9.20}$$

while the data rate at user 2 can be written as

$$\mathcal{I}_2 = \min(R_{1\to r}, R_{r\to 2}) \tag{9.21}$$

$$= \min\left(\frac{1}{4}\sum_{k=1}^{M}\log_2\left(1+\frac{a_k\lambda_{1,k}}{\sigma_r^2}\right), \frac{1}{4}\sum_{k=1}^{M}\log_2\left(1+\frac{d_k\lambda_{2,k}}{\sigma_2^2}\right)\right). \tag{9.22}$$

Transmission power is allocated equally among nodes and data streams. Recall that in the previous subsection, the network is subject to a total power constraint of P, for every two time slots. Therefore, the power allocation factors are defined as $a_k = \frac{P}{2M}$, $b_k = \frac{P}{2M}$, $c_k = \frac{P}{2M}$, and $d_k = \frac{P}{2M}$, $\forall k = \{1, \ldots, M\}$.

Comparable Scheme: Alternate Optimization (A-Opt)

The best comparable scheme for the non-regenerative MIMO two-way relaying channel is the A-Opt scheme proposed in Xu and Hua [65]. Due to the fact that the sum-rate expression computable from (9.3) is non-concave, Xu and Hua [65] propose the A-Opt scheme, which alternately computes locally optimal source beamformers for fixed relay beamformer and locally optimal relay beamformer for fixed source beamformers until convergence is reached. Several searching algorithms are proposed in Xu and Hua [65] to determine locally optimal beamforming matrices subject to individual power constraints and assuming the use of perfect MIMO decoders at the users. Although the A-Opt scheme is able to achieve the best sum-rate under individual power constraints and symmetric SNR, it is computationally expensive to determine the beamforming matrices. Generally, the problem dimension of the searching algorithms in Xu and Hua [65] grows quadratically with the number of antennas. This increases the computational complexity significantly when a higher number of antennas is used. In comparison, the problem dimension of the proposed JPA I and II in this chapter is linear with the number of antennas. Furthermore, due to the fact that the sum-rate is non-concave for any fixed source beamformers, the searching algorithms in Xu and Hua [65] have to be repeated multiple times with different starting points in order to increase the probability of finding the global optimal relay beamformer. This further increases the computational overhead.

In order to obtain the simulation results for A-Opt, the weighted minimum MSE algorithm proposed in Xu and Hua [65] is used to compute the relay beamforming matrix, while the semi-definite program solver in CVX toolbox [71, 72] is used to compute the user beamforming matrices. Each node is subject to individual power constraint $\frac{P}{3}$, which sums up to a joint power constraint P, to enable fair comparison with the proposed strategies in this chapter.

9.5 NUMERICAL RESULTS

This section presents the numerical results of the proposed joint beamforming and power management scheme in comparison with existing schemes. The optimization problems discussed in the previous sections are solved using the nonlinear optimization toolbox in MATLAB (i.e., using the function $fmincon$ and the interior-point method). The ergodic sum-rates of various schemes with fixed total power constraint are simulated using the Monte Carlo method. The relationship between the ergodic sum-rate and parameters such as SNR, number of antennas, and path loss are discussed in the following paragraphs.

Figure 9.4 shows the ergodic sum-rate versus reference SNR ($\frac{1}{\sigma_1^2} = \frac{1}{\sigma_2^2} = \frac{1}{\sigma_r^2}$) of the proposed JPA schemes in comparison with existing schemes. The reference SNR is defined as the inverse of the noise power. In the subsequent discussion, the term SNR is used to imply the reference SNR defined here. In this simulation, the SNRs at all nodes are assumed to be symmetrical (equal noise power), that is, $\frac{1}{\sigma_1^2} = \frac{1}{\sigma_2^2} = \frac{1}{\sigma_r^2}$. The fixed parameters are the number of antennas, $M = 4$ and the total power constraints, $P = 3$ watts. From the figure, it can be observed that the proposed JPA I and II perform close to the A-Opt scheme at high SNR. In the range of low to medium SNR, the performance gain contributed by the proposed JPA schemes over baseline pure AF scheme is limited if compared to the A-Opt scheme. This is due to the

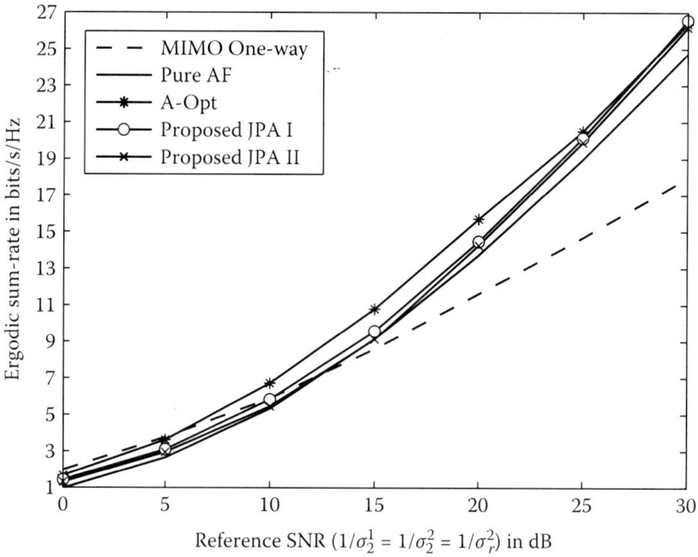

FIGURE 9.4 Ergodic sum-rate versus SNR ($\frac{1}{\sigma_1^2} = \frac{1}{\sigma_2^2} = \frac{1}{\sigma_r^2}$) for fixed $M = 4$, $P = 3$.

fact that the choice of beamforming directions in the proposed JPA schemes is suboptimal if compared to the A-Opt scheme. However, at high SNR, the suboptimal beamforming directions do not prevent the proposed JPA schemes from delivering significant performance gain against the pure AF scheme through dynamic allocation of power among substreams and nodes. It can be observed also that the JPA II performs close to the proposed JPA I. This indicates that the upper bound in Theorem 9.4.2 is a good approximation of the original problem in (9.16). The performance gaps between two-way relaying schemes (Pure AF, A-Opt, proposed JPA I and II) and one-way relaying scheme enlarge with the increase of SNR. It is evident from the slope of the sum-rate curves that, two-way relaying schemes are able to achieve higher multiplexing gain due to more efficient use of bandwidth.

In the following simulations, the ergodic sum-rates of the various schemes when the SNR at users and relay are asymmetrical (unequal noise power) are simulated. Figure 9.5 shows the ergodic sum-rate versus SNR at the users ($\frac{1}{\sigma_1^2} = \frac{1}{\sigma_2^2}$) when SNR at the relay is fixed at $\frac{1}{\sigma_r^2} = 30$ dB. Other fixed parameters are $M = 4$ and $P = 3$ watts. From the figure, it can be observed that the proposed JPA I achieves the best ergodic sum-rate, followed closely

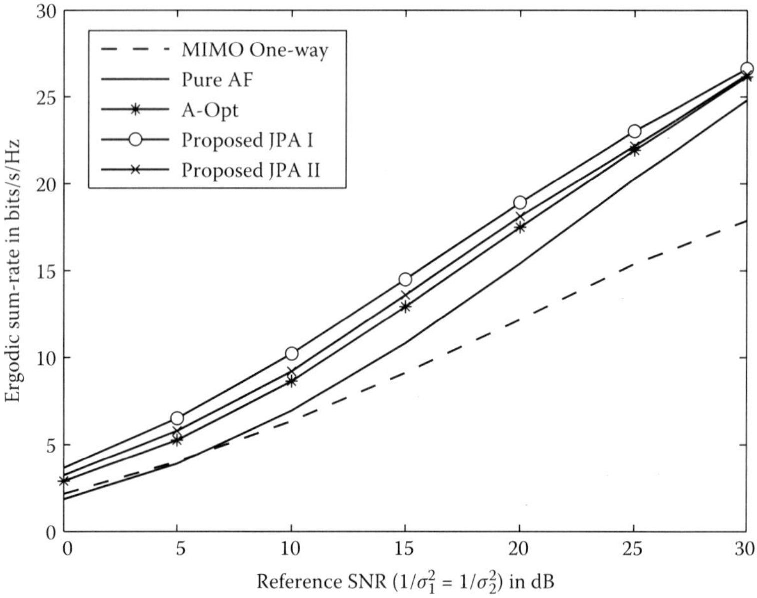

FIGURE 9.5 Ergodic sum-rate versus SNR ($\frac{1}{\sigma_1^2} = \frac{1}{\sigma_2^2}$) for fixed $M = 4$, $P = 3$, and $\frac{1}{\sigma_r^2} = 30$ dB.

by the proposed JPA II. The A-Opt scheme does not perform better than the proposed JPA schemes. This is due to the fact that, in the A-Opt scheme, each node is allocated with a fixed amount of power that does not correlate with the asymmetric SNR. In comparison, the proposed JPA schemes respond to the asymmetric SNR by allocating power dynamically among nodes and substreams. The joint power allocation between nodes provides another dimension of improvement. Figure 9.6 shows the ergodic sum-rate versus SNR at the relay ($\frac{1}{\sigma_r^2}$) when the SNRs at the users are fixed at $\frac{1}{\sigma_1^2} = \frac{1}{\sigma_2^2} = 5$ dB. Other fixed parameters are $M = 4$ and $P = 3$ watts. From the figure, it is clear that the proposed JPA schemes deliver significant performance gain over the pure AF scheme. The proposed JPA I and JPA II deliver higher ergodic sum-rate than the A-Opt scheme when SNR is greater than 15 dB and 20 dB, respectively. This supports that dynamic power allocation between users and relay is able to utilize the asymmetric SNR between nodes to obtain better sum-rate performance. At low SNR, the proposed schemes do not perform as well as the A-Opt scheme due to the use of suboptimal beamforming directions. It is interesting to observe that at low SNR, the baseline MIMO one-way relaying

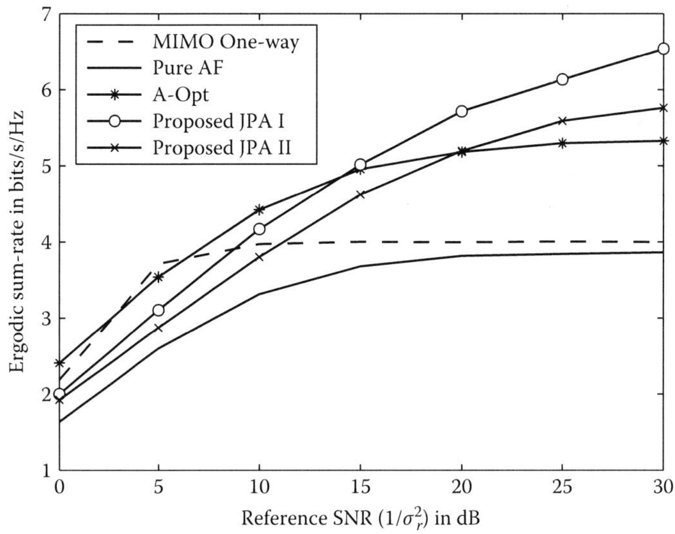

FIGURE 9.6 Ergodic sum-rate versus SNR $(\frac{1}{\sigma_r^2})$ for fixed $M = 4$, $P = 3$, and $\frac{1}{\sigma_1^2} - \frac{1}{\sigma_2^2} - 5$ dB.

performs as well as the A-Opt scheme. At low SNR, the performance of two-way relaying schemes (Pure AF, A-Opt, proposed JPA I and II) is limited by not only the noise at the users but also the propagated noise from the relay. As a result, two-way relaying schemes do not perform better than one-way relaying schemes at low SNR. This effect can also be observed in Figure 9.4.

Figure 9.7 shows the ergodic sum-rate versus number of antennas M of various schemes. The fixed parameters are $\frac{1}{\sigma_1^2} = \frac{1}{\sigma_2^2} = 15$ dB, $\frac{1}{\sigma_r^2} = 30$ dB, and $P = 3$ watts. Generally, the ergodic sum-rates of all schemes increase linearly with the number of antennas, M. As the numbers of antennas at all nodes are increased simultaneously, the number of independent data streams supportable in the network increases. In other words, the multiplexing gain grows linearly with M. Among all power allocation schemes, the proposed JPA I achieves the best sum-rate performance, followed closely by the proposed JPA II. The proposed JPA schemes outperform the A-Opt scheme, thanks to the joint power allocation between nodes. From the figure, it can be observed that the gaps between proposed schemes and pure AF scheme enlarge for increasing M. This shows that joint power allocation is vital in delivering better data rates in system with high multiplexing gain.

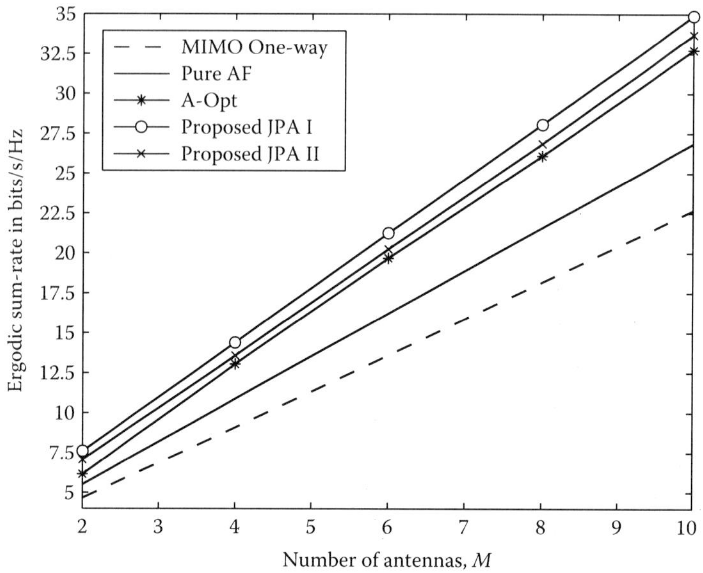

FIGURE 9.7 Ergodic sum-rate versus number of antennas M, for fixed $\frac{1}{\sigma_1^2} = \frac{1}{\sigma_2^2} = 15$ dB, $\frac{1}{\sigma_r^2} = 30$ dB, and $P = 3$.

In the next simulation, the effect of large-scale path loss to the ergodic sum-rate is investigated. The path loss is integrated in the channel model as $1/\sqrt{d_i^\alpha}\mathbf{H}_i$, where d_i is the distance between user i and the relay, α is the path-loss exponent, and $\mathbf{H}_i$ is the channel matrix between user i and the relay (as shown in Section 9.2) where its entries are i.i.d. Rayleigh distributed. A simple line network is considered where the relay is placed in between the users. Figure 9.8 shows the ergodic sum-rate versus the relay location d_r of various schemes. The distance between user 1 and the relay is $d_1 = 1 + d_r$, while the distance between user 2 and the relay is $d_2 = 2 - d_r$. The constant offset in d_i is introduced in order to ensure that the received power does not exceed the transmitted power. The fixed parameters are $M = 2$, $\alpha = 4$, $\frac{1}{\sigma_1^2} = \frac{1}{\sigma_2^2} = \frac{1}{\sigma_r^2} = 30$ dB, and $P = 3$ watts. In general, all schemes achieve their best sum-rate when the relay is located in the middle of the users. The proposed JPA I delivers the best sum-rate performance and has the least sensitivity toward the variation of the location of the relay. The proposed JPA II performs close to JPA I, but it is more sensitive to unequal path loss due to the fact that both users are allocated equal amount of power. Refer to

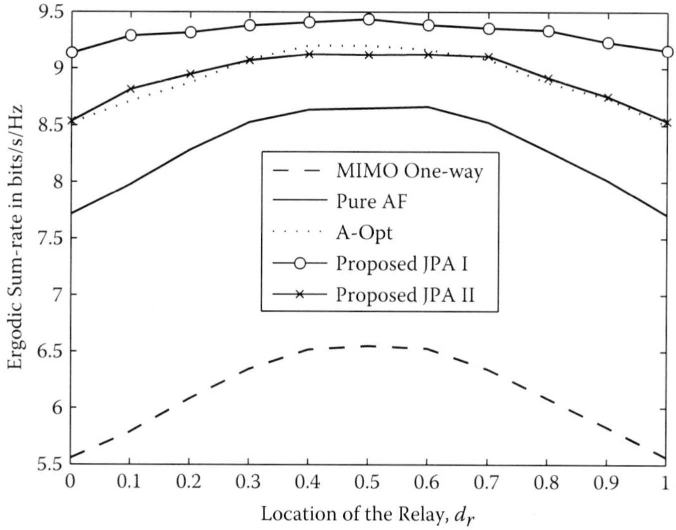

FIGURE 9.8 Ergodic sum-rate versus the relay location d_r for fixed $M = 2$, $\frac{1}{\sigma_1^2} = \frac{1}{\sigma_2^2} = \frac{1}{\sigma_r^2} = 30$ dB, $P = 3$, $\alpha = 4$.

the remark of Subsection 9.4.2. The A-Opt scheme performs similarly to the proposed JPA II. The baseline MIMO one-way relaying scheme displays the worst sum-rate performance and the highest sensitivity toward unequal path loss.

9.6 CONCLUSION

Joint beamforming and power allocation in the MIMO two-way relaying channel with a non-regenerative relay has been proposed. The proposed beamformers use the subchannel alignment technique to decompose the user channel pair into parallel subchannels. The proposed joint power allocation maximizes the sum-rate subject to the network power constraint. The non-concave power allocation utility function is approximated by a concave upper bound to facilitate the use of efficient convex optimization techniques in solving the optimization problems. Numerical simulation results demonstrate that the proposed joint beamforming and power allocation scheme is able to deliver significant sum-rate improvement when compared to existing schemes.

CHAPTER 10

Selfishness-Aware Energy-Efficient Cooperative Networks

Jun Fan, Beijing Institute of Technology, China
Zhengguo Sheng, France Telecom Orange Labs, China
Chi Harold Liu, Beijing Institute of Technology, China

10.1 INTRODUCTION

In wireless communication, signal fading caused by multi-path propagation is a particularly serious channel detriment. In order to address this issue, cooperative communication mechanisms [1, 73, 74] have been proposed as an effective way of exploiting spatial diversity to improve the quality of wireless links. By encouraging single-antenna devices to share their antennas cooperatively, a number of potential benefits can be achieved, including improved reception reliability [75], reduced power consumption [76], and increased spectrum efficiency [77].

Although the idea of cooperative communication has been proposed for almost a decade, there are still fundamental issues to be considered from theoretical analysis perspectives; in particular, we focus on the fundamental issue of relay selection in this chapter. In the existing literature, a number of schemes have been proposed, ranging from single relay selection [24, 78] to multi-relay selection [79, 80, 81]. However, none of these works considers the fair usage of energy consumption, especially for relay nodes. Motivated by the recent research on the study of the inherent loss of efficiency caused to a system by the participant's selfishness [82], we explicitly consider the impact of selfish behavior on relay selection and uniquely propose an effective mechanism to achieve energy fairness.

More specifically, in this chapter, we propose a cooperation scheme to jointly consider relay selection and power allocation by incorporating fairness and selfish natures of wireless nodes. We capture the selfish behavior in wireless networks by introducing a *selfishness index*, which represents the time-varying selfish level of each node, and incorporate this index into a novel *utility* function, which denotes the net payoff of a node from cooperative transmission. Higher utility values denote more responsibility in the cooperative transmission; for the purpose of decreasing power consumption, a node with lower utility can show some degree of selfishness to reserve energy.

The contribution of the chapter is threefold:

- We explicitly derive a closed-form solution for the outage probability of a multi-relay cooperation scheme employing repetition-coded decode-and-forward (DF) strategy.

- We incorporate a novel concept of *selfishness index* into a utility function to uniquely capture and regulate the selfish behavior of a node in the proposed relay selection strategy.

- We propose a two-step relay selection mechanism covering all aspects of power efficiency, energy consumption fairness, and network lifetime.

The remainder of this chapter is organized as follows. In Section 10.2, we present the system model, derive the closed-form expression of outage probability for multiple simultaneous relays under repetition-coded DF cooperation, and define the utility parameter with the selfishness index. The proposed approach of relay selection and power allocation is introduced in Section 10.3 and 10.4. Simulation results are shown in Section 10.5, followed by concluding remarks in Section 10.6.

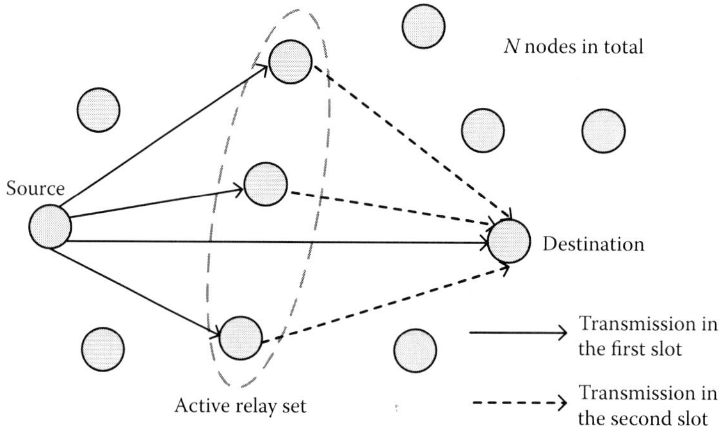

FIGURE 10.1 Cooperative transmission model with multiple relays.

10.2 SYSTEM MODEL

We consider a wireless cooperative relay network $\mathcal{N}$ consisting of N nodes, where $\mathcal{N} \triangleq \{i = 1, 2, \ldots, N\}$. All nodes possess the same initial energy reserve, denoted as E_i, and the residual energy at time t is denoted as $\bar{E}_i(t), \forall i \in \mathcal{N}$. We assume a discrete system operation where time is divided into discrete time slots, and TDMA is used to provide collision-free transmissions from the source and relays [1]. As in Chen et al. [80], we assume full channel state information is available at the source, where centralized relay selection and power allocation is performed. We assume that relay transmissions are half-duplex such that they cannot send and receive packets at the same time. All channels exhibit additive white Gaussian noise (AWGN). Finally, the fading is assumed to be stationary, with frequency non-selective Rayleigh block fading.

As shown in Figure 10.1, we conduct an investigation into the two-time-slot implementation of repetition-coded DF relaying strategies. At time t, a source s intends to send a message to a destination d. Other nodes are considered to be the candidate relays. In the first time slot, the source selects K nodes from the candidate relays to form an active relay set $\mathcal{K}_s$. Then it broadcasts its packets to both the active relays and the destination. In the second time slot, the active relays in $\mathcal{K}_s$ retransmit the received packets, operating in a perfect synchronous manner to obtain the "emergence" diversity gain [80]. Hence, the destination receives multiple independent copies of the same packets transmitted through various wireless channels, and cooperative diversity gain can be achieved.

10.2.1 Direct Transmission

We start with direct transmission. According to the system model assumptions, the channel model incorporates path loss and Rayleigh fading. The channel gain $\mathrm{a}_{s,d}$ between the nodes s and d is modeled as $\mathrm{a}_{s,d} = h_{s,d}/d_{s,d}^{\alpha/2}$, where $d_{s,d}$ is the Euclidian distance between the nodes s and d, α is the path-loss exponent,[1] and $h_{s,d}$ captures the channel fading characteristics.

The mutual information of the cooperative link is a random variable denoted by I. For a target rate R, $I < R$ represents the outage events, and we use $\epsilon^{\text{out}} \triangleq \Pr[I < R]$ to denote its outage probability [1]. Then, the mutual information between source s and destination d is

$$I_{s,d}^{\text{DT}} = \log(1 + P_s^D|\mathrm{a}_{s,d}|^2), \quad \forall s, d \in \mathcal{N}, \tag{10.1}$$

where P_s^D is the normalized transmission power of source s, and for Rayleigh fading, $|\mathrm{a}_{s,d}|^2$ is exponentially distributed with parameter $d_{s,d}^\alpha$. Therefore, the outage probability satisfies

$$\begin{aligned} \epsilon^{\text{out}} &= \Pr[I_{s,d}^{\text{DT}} < R] = \Pr\left[|\mathrm{a}_{s,d}|^2 < \frac{2^R - 1}{P_s^D}\right] \\ &= 1 - \exp\left(-\frac{(2^R - 1)d_{s,d}^\alpha}{P_s^D}\right) \approx d_{s,d}^\alpha\left(\frac{2^R - 1}{P_s^D}\right), \end{aligned} \tag{10.2}$$

under the condition that P_s^D is large and R is the desired data rate in bit/s/Hz, which can be defined by specific QoS requirements (e.g., routing demand).

Then, we write the normalized transmission power for direct transmission as

$$P_s^D = d_{s,d}^\alpha\left(\frac{2^R - 1}{\epsilon^{\text{out}}}\right). \tag{10.3}$$

[1] The path-loss exponent α is experimentally determined and is typically in the range of 2 to 5 depending on propagation environment. For example, $\alpha = 2.0$ is for free space, 2.5–3.0 for rural areas, 3.0–4.0 for urban areas, and 4.0–5.0 for dense urban areas.

10.2.2 Cooperative Transmission with Multiple Simultaneous Relays

For a given active relay set $\mathcal{K}_s$ with size K, the maximum average mutual information between source s and destination d for repetition-coded DF [1, 83] can be shown as

$$I_{s,d}^{\mathrm{CT}} = \min\{I_{s,\mathcal{K}_s}, I_{\mathcal{K}_s,d}\}, \tag{10.4}$$

where $I_{s,\mathcal{K}_s}$ and $I_{\mathcal{K}_s,d}$ are the mutual information in the first and second time slot, respectively, defined as

$$\begin{aligned} I_{s,\mathcal{K}_s} &= \frac{1}{2}\log\left(1 + P_s^C|\mathrm{a}_{s,\mathcal{K}_s}|^2\right), \\ I_{\mathcal{K}_s,d} &= \frac{1}{2}\log\left(1 + P_s^C|\mathrm{a}_{s,d}|^2 + \sum_{i\in\mathcal{K}_s} P_i^C|\mathrm{a}_{i,d}|^2\right), \end{aligned} \tag{10.5}$$

where P_s^C is the normalized power of source s in cooperative transmission, P_i^C is the normalized power of active relay $i \in \mathcal{K}_s$, and $\mathrm{a}_{s,\mathcal{K}_s} = 1/K\sum_{i\in\mathcal{K}_s} \mathrm{a}_{s,i}$ is the average source-relay channel gain. Factor 1/2 indicates the fact that communication is performed in two time slots.

To simplify the discussion, each of the selected relays working under the DF protocol is assumed to have enough signal-to-noise ratio (SNR) to decode the received signals successfully [80]. Based on this, we can conclude that the mutual information bottleneck between s and d is the second term in (10.5), namely $I_{\mathcal{K}_s,d}$. The following analysis is under the condition of $I_{s,K_s} > I_{K_s,d}$. Then, we formulate the outage event given by $I_{s,d}^{\mathrm{CT}} < R$ and the outage probability, as

$$\begin{aligned} \epsilon^{\mathrm{out}} &= \Pr[I_{s,d}^{\mathrm{CT}} < R] = \Pr[I_{\mathcal{K}_s,d} < R] \\ &= \Pr\left[P_s^C|\mathrm{a}_{s,d}|^2 + \sum_{i\in\mathcal{K}_s} P_i^C|\mathrm{a}_{i,d}|^2 < 2^{2R} - 1\right]. \end{aligned} \tag{10.6}$$

Theorem 10.1 *For multiple simultaneous relays, the outage probability under repetition-coded DF cooperation is given by*

$$\begin{aligned} \epsilon^{\mathrm{out}} &= Pr\left[P_s^C|a_{s,d}|^2 + \sum_{i\in\mathcal{K}_s} P_i^C|a_{i,d}|^2 < 2^{2R} - 1\right] \\ &\approx \frac{(2^{2R}-1)^{K+1}}{(K+1)!}\prod_{i\in\{s\}\cup\mathcal{K}_s}\frac{d_{i,d}^{\alpha}}{P_i^C}. \end{aligned} \tag{10.7}$$

Proof Laneman [84] proposed an approximation for (10.6) but with the assumption that each node has the same normalized power. We extend this result to the scenario where the normalized power of each node P_i^C, $\forall i \in \{s\} \cup \mathcal{K}_s$ can be different. Take one item from (10.7) and let $\delta = 2^{2R} - 1$, $\rho_i s = P_i^C$; then we have

$$\begin{aligned}\lim_{s\to\infty} s \cdot \Pr[P_i^C |\mathrm{a}_{i,d}|^2 < \delta] &= \lim_{s\to\infty} s \cdot \Pr[\rho_i s |\mathrm{a}_{i,d}|^2 < \delta] \\ &= \lim_{s\to\infty} s \cdot \Pr\left[|\mathrm{a}_{i,d}|^2 < \frac{\delta}{\rho_i s}\right] \\ &= \lim_{s\to\infty} s \cdot \frac{\delta}{\rho_i s} d_{i,d}^{\alpha} = \frac{\delta}{\rho_i} d_{i,d}^{\alpha}.\end{aligned} \tag{10.8}$$

By applying Theorem 10.10.1 in Laneman [84], the result (10.8) being utilized K times, yields the approximation

$$\begin{aligned}\epsilon^{\text{out}} &= \Pr\left[P_s^C |\mathrm{a}_{s,d}|^2 + \sum_{i\in\mathcal{K}_s} P_i^C |\mathrm{a}_{i,d}|^2 < 2^{2R} - 1\right] \\ &\approx \frac{(2^{2R}-1)^{K+1}}{(K+1)!} \prod_{i\in\{s\}\cup\mathcal{K}_s} \frac{d_{i,d}^{\alpha}}{P_i^C}.\end{aligned} \tag{10.9}$$

□

From (10.7), it is clear that the outage probability is closely related to the transmission power P_i^C, number of selected relays K, desired data rate R, path-loss exponent α, and the distance between nodes $d_{i,d}$, $\forall i \in \{s\} \cup \mathcal{K}_s$.

10.2.3 Utility Function and Selfishness Index

To achieve the fairness in relay selection and power allocation, we associate each node with a utility representing its net payoff (or actual benefit) received from cooperative transmission. Existing literature has shown that by using relaying, a source node can gain a certain degree of power savings. However, if a node serves as a relay, it contributes its own resources to help others. Given the specific employed MAC scheduling algorithm, some nodes may act more as the source while others serve more as the relay. Therefore by introducing the utility, we aim to balance the power savings and loss of each node and offer a certain degree of fairness.

We use parameter *gain* G to represent the power savings of a source from cooperative communication. The instantaneous gain of node i at time t is defined as the power difference between using direct transmission and cooperative relaying:

$$G_i(t) = G_i(t-1) + \Delta G_i(t), \tag{10.10}$$

and

$$\Delta G_i(t) = \begin{cases} P_i^D(t) - P_i^C(t), & i = s, \\ 0, & \text{others.} \end{cases} \tag{10.11}$$

Next, we introduce a parameter *loss* L representing the power spending of a relay from cooperative communication. The instantaneous loss of node i at time t is defined as

$$L_i(t) = L_i(t-1) + \Delta L_i(t), \tag{10.12}$$

and

$$\Delta L_i(t) = \begin{cases} P_i^C, & \forall i \in \mathcal{K}_s, \\ 0, & \text{others.} \end{cases} \tag{10.13}$$

At time t, only relays can accumulate a certain amount of loss as they are helpers to the source and make a sacrifice in relaying packets. Since gain and loss are time-accumulative parameters, it is obvious that both $G_i(0)$ and $L_i(0)$ equal zero, and if a node does not transmit any packets at time t, its gain and loss remain unchanged.

We use utility $U_i(t)$ to denote the actual benefit node i retains at time t. One straightforward method to define a utility is to subtract the loss from the gain. Meanwhile, considering the selfish behavior of nodes, we incorporate a selfishness index $\gamma_i(t) \in [0, 1]$ into the utility function to make it more rational. Then, the utility $U_i(t)$ of each node i at time t can be formulated as

$$U_i(t) = \gamma_i(t)\left(G_i(t) - L_i(t)\right), \forall i \in \mathcal{N}. \tag{10.14}$$

It is worth noting that γ can be arbitrarily chosen where value "0" represents extremely selfish behavior and "1" indicates highly generous. Considering the residual energy of a node at any time, we further quantify γ as

$$\gamma_i(t) = \frac{\bar{E}_i(t)}{E_i}, \quad \forall i \in \mathcal{N}. \tag{10.15}$$

We argue that selfish behaviors can be fully captured in parameter γ and utility function (10.14). First, if we denote the difference between the gain and loss as the *primal* utility, then "selfishness" means that a node is not willing to cooperate so as to take less responsibility in relay selection and power consumption. Examining (10.14), we see that the time-varying selfishness index $\gamma_i(t)$ reflects the percentage of utility (or net payoff) discounted at each node. Given two nodes with the same primal utility, it is obvious that the one with less residual energy would have a lower utility and thus take less responsibility in helping others.

We associate the utility (actual benefit) with transmission "responsibility" in two folds: one is in relay selection process, where only nodes with nonnegative utility have the opportunity to be selected as active relays; the second is in power allocation process, where the power of a node is proportional to its utility. In the next section, we will show how the utility performs in the proposed optimal power allocation.

10.3 OPTIMAL POWER ALLOCATION

Since the proposed power allocation rule is highly related to the relay selection criteria, we start with the assumption that the active relay set $\mathcal{K}_s$ is given, and in the next section, we illustrate the mechanism to find the most appropriate K relays.

At time t, for given K relays, we propose an optimal power allocation scheme as a constrained optimization problem aiming at minimizing the sum of weighted transmission power under a given outage probability threshold ϵ_0. The weight of node i is defined as

$$w_i(t) = g(\dot{U}_i(t)), \quad \forall i \in \mathcal{N}, \tag{10.16}$$

where $g(\dot{U})$ is a generic non-increasing function of the normalized utility value $\dot{U}$,[2] and one of its realizations can be $g(\dot{U}) = e^{(-\beta\dot{U})}$, where β is a constant.

Now we formally introduce the optimization problem as

$$\begin{aligned} \{P_i^C(t)\}_{i\in\{s\}\cup\mathcal{K}_s} &= \underset{P_i^C(t)}{\operatorname{argmin}} \sum_{i\in\{s\}\cup\mathcal{K}_s} w_i(t)P_i^C(t), \\ \text{s.t.} \quad \epsilon^{\text{out}} &\le \epsilon_0. \end{aligned} \tag{10.17}$$

[2]In order to derive a reasonable result, we scale the utility U into a value range of [0,1].

Theorem 10.2 *The optimal transmission power for the proposed relay selection scheme, given that a target QoS is satisfied, is given by*

$$P_i^C(t) = \sqrt[K+1]{\frac{Q}{\epsilon_0}} \cdot \frac{\sqrt[K+1]{\prod_{i\in\{s\}\cup\mathcal{K}_s} w_i(t)}}{w_i(t)}, \quad \forall i \in \{s\} \cup \mathcal{K}_s, \tag{10.18}$$

where $Q = \frac{(2^{2R}-1)^{K+1}}{(K+1)!} \prod_{i\in\{s\}\cup\mathcal{K}_s} d_{i,d}^{\alpha}$.

Proof According to Theorem 10.1, let Q be a constant

$$Q = \frac{(2^{2R}-1)^{K+1}}{(K+1)!} \prod_{i\in\{s\}\cup\mathcal{K}_s} d_{i,d}^{\alpha},$$

given R, K, and $\mathcal{K}_s$; then ϵ^{out} can be rewritten as

$$\epsilon^{\text{out}} = Q \Big/ \left[\prod_{i\in\{s\}\cup\mathcal{K}_s} P_i^C(t)\right].$$

Hence, the optimization problem (10.17) becomes

$$\begin{aligned} \{P_i^C(t)\}_{i\in\{s\}\cup\mathcal{K}_s} &= \underset{P_i^C(t)}{\operatorname{argmin}} \sum_{i\in\{s\}\cup\mathcal{K}_s} w_i(t) P_i^C(t), \\ \text{s.t.} \quad \prod_{i\in\{s\}\cup\mathcal{K}_s} P_i^C(t) &\geq \frac{Q}{\epsilon_0}. \end{aligned} \tag{10.19}$$

It is a classic constrained optimization problem that could be solved by Lagrangian multipliers, and we form a Lagrangian problem with multiplier λ as

$$F = \sum_{i\in\{s\}\cup\mathcal{K}_s} w_i(t) P_i^C(t) - \lambda \left(\prod_{i\in\{s\}\cup\mathcal{K}_s} P_i^C(t) - \frac{Q}{\epsilon_0}\right). \tag{10.20}$$

The first-order (necessary) optimality condition for (10.20) is

$$\nabla F = 0 \text{ and } \lambda \left(\prod_{i\in\{s\}\cup\mathcal{K}_s} P_i^C(t) - \frac{Q}{\epsilon_0}\right) = 0. \tag{10.21}$$

Since the constraint is binding, $\lambda \neq 0$ and $\prod_{i\in\{s\}\cup\mathcal{K}_s} P_i^C(t) - \frac{Q}{\epsilon_0} = 0$, the first part of (10.21) becomes

$$\nabla F = 0 \Leftrightarrow P_i^C(t)w_i(t) = \frac{\lambda Q}{\epsilon_0}, \quad \forall i \in \{s\} \cup \mathcal{K}_s, \tag{10.22}$$

or equivalently,

$$P_i^C(t) = \frac{\lambda Q}{\epsilon_0 w_i(t)}, \quad \forall i \in \{s\} \cup \mathcal{K}_s. \tag{10.23}$$

In other words, the instantaneous power consumption of node i is inversely proportional to its weight $w_i(t)$. It is worth noting that from (10.16) we define $w_i(t)$ as a non-increasing function of the normalized utility $\dot{U}_i(t)$, and more importantly, it can be concluded that the optimal power consumption for node i is proportional to its instantaneous utility.

Next, we put (10.23) into the constraint $\prod_{i\in\{s\}\cup\mathcal{K}_s} P_i^C(t) = Q/\epsilon_0$ and have

$$\lambda = \sqrt[K+1]{\frac{\prod_{i\in\{s\}\cup\mathcal{K}_s} w_i(t)}{(Q/\epsilon_0)^K}}. \tag{10.24}$$

Putting the above result into (10.23) leads to the final result. □

From (10.18), it is clear that higher utility values result in higher power allocation. If a node benefits more from cooperation (acting as a source most of time) or has more residual energy, it should take more responsibility on the transmission. Those who contribute a lot as relays or have less energy left can exhibit different degrees of selfishness to preserve energy. Hence, by minimizing the sum of weighted power of each node, optimal power efficiency and a certain degree of energy consumption fairness can be jointly achieved.

10.4 NETWORK LIFETIME-AWARE TWO-STEP RELAY SELECTION

This section aims to find the most appropriate K nodes to form relay set $\mathcal{K}_s$, given that the transmission power of these K relays can already be optimally allocated as (10.18). The selection mechanism should allow the selected relays to achieve a high degree of power efficiency. Furthermore, as the benefit of each node receiving from the cooperative transmission is different, to be fair and equitable, nodes with negative benefit from cooperative transmission, that is, $U_i(t) < 0$, would be removed from relay selection. Moreover, since most terminals are battery-powered, it should balance the energy consumption of each node as much as possible to maximize the network lifetime, which, in this chapter, is defined as when the first node runs out of energy. In the following discussions, we illustrate how power efficiency, cumulative benefits, energy consumption fairness, and network lifetime are considered in relay selection.

At time t, we denote the nodes with nonnegative utility as a set $\mathcal{M} \triangleq \{m = 1, 2, \ldots, M\}$, where $U_m(t) \in \mathbb{R}^+, \forall m \in \mathcal{M}$. Then, the active relay set $\mathcal{K}_s$ with variable size K is a subset of $\mathcal{M}$, and $K = 1, 2, \ldots, M$. To this end, by performing an exhaustive search over all possible decoding sets, we have $\sum_{K=1}^{M} \binom{M}{K}$ different combinations, or optimization iterations. We next separate this exhaustive search into two steps. Step 1 takes power efficiency and energy consumption fairness into consideration by using (10.18) to optimally allocate transmission power, and Step 2 covers the aspect of network lifetime.

Step 1: We choose the most appropriate relay set for each $K = 1, 2, \ldots, M$, and for the totally $\binom{M}{K}$ potential options, we allocate the power as (10.18) and find the set that minimizes the objective function, the result of which is denoted as $\mathcal{K}^* \mid_K$,

$$\begin{aligned} &\mathcal{K}^* \mid_K = \underset{\mathcal{K}^j \mid_K}{\operatorname{argmin}} \sum_i w_i(t) P_i^C(t), \\ &\forall i \in \{s\} \cup \mathcal{K}^j \mid_K, \quad \forall j \in \left\{1, 2, \ldots, \binom{M}{k}\right\}, \end{aligned} \tag{10.25}$$

where $\mathcal{K}^j \mid_K$ represents the *j*th subset of $\mathcal{M}$ with its size $|\mathcal{K}^j| = K$. Then, we calculate the estimated network lifetime T of set $\mathcal{K}^*|_K$ as

$$T(\mathcal{K}^* \mid_K) = \min_i \frac{\bar{E}_i(t)}{P_i^C(t)}, \quad \forall i \in \{s\} \cup \mathcal{K}^* \mid_K . \tag{10.26}$$

Step 2: From the previous obtained M optional sets, we select the one with the longest estimated network lifetime as the active relay set $\mathcal{K}_s$:

$$\mathcal{K}_s = \underset{\mathcal{K}^* \mid_K}{\operatorname{argmax}}\, T, \quad \forall K \in \{1, 2, \ldots, M\}. \tag{10.27}$$

ALGORITHM 10.1 Power Allocation and Relay Selection

1 $\mathcal{M} = \varnothing$;
2 **foreach** $i \in \mathcal{N}$ **do**
3   **if** $i \neq s$ *and* $i \neq d$ *and* $U_i(t) \in \mathbb{R}^+$ **then**
4         $\mathcal{M} = \mathcal{M} \cup \{i\}$;
5   **end**
6 **end**
7 $M = |\mathcal{M}|$;
8 **for** $k = 1$ **to** M **do**
9   **for** $j = 1$ **to** $\binom{M}{k}$ **do**
10   set $\mathcal{K}^j\ |_K$ as the *j*th subset of $\mathcal{M}$ with size K;
11   allocate the power of each node i as (10.18),$\forall i \in \{s\} \cup \mathcal{K}^j\ |_K$;
12   **end**
13   calculate $\mathcal{K}^*\ |_K$ as in (10.25);
14   calculate $T(\mathcal{K}^*\ |_K)$ as in (10.26);
15 **end**
16 $\mathcal{K}_s = \text{argmax}_{\mathcal{K}^*|_K} T, \forall K \in \{1, 2, \ldots, M\}$;
17 allocate the power of each node i as (10.18), $\forall i \in \{s\} \cup \mathcal{K}_s$.

Algorithm 10.1 shows the pseudo-code of our proposed power allocation and relay selection algorithm. At time t, given the source s, destination d, and the candidate relays, we run this algorithm to generate the active relay set $\mathcal{K}_s$ and allocate the power of source and active relays. From lines 1–5, we select the relays with nonnegative utility to form the set $\mathcal{M}$. Lines 8–15 show the exhaustive search over all possible decoding sets, where the internal iteration is the Step 1 of relay selection and external iteration is the Step 2 of relay selection, as stated above. Finally, we get the active relay set $\mathcal{K}_s$ at line 16 and allocate the power of each node at line 17.

10.5 PERFORMANCE EVALUATION

In this section, we use two different network settings to investigate the impact of selfish behavior on power consumption and provide extensive numerical results to demonstrate the superiority of our proposed approach in power savings and energy consumption fairness.

10.5.1 A Five-Node Example

In this scenario, we randomly deploy five nodes in a 100×100 m^2 region (the edges of the region are wrapped (toroid) to eliminate edge effects).

Throughout the simulation, we set the path-loss exponent $\alpha = 3$, data rate $R = 1$ bps/Hz, and the targeted outage $\epsilon_0 = 0.01$, all as constants. Furthermore, we use $g(\dot{U}) = e^{-\dot{U}}$. At each time t, a packet is transmitted by a randomly selected source and received by a randomly selected destination. The other three nodes are treated as candidate relays from which the source generates the active relay set $\mathcal{K}_s$. A total of 1000 packets are transmitted within the network, and all nodes have adequate energy to be alive during the simulation.

We run our proposed approach in four different parameter settings. In the fully generous setting (dark gray), none of the nodes behave selfishly at any time, that is, $\gamma_i(t) = 1, \forall i, t$. In the second (light gray) and third (medium gray) settings, we partially assign nodes 3 and 5 as selfish nodes with $\gamma_{3,5}(t) = 0.5, 0.1, \forall t$, respectively, while for other nodes we set $\gamma_{1,2,4}(t) = 1, \forall t$. Finally, in the all energy-aware selfish setting (black), all nodes exhibit time-varying selfish behavior proportional to the percentage of remaining energy $\gamma_i(t) = \bar{E}_i(t)/E_i, \forall i, t$.

In Figure 10.2(a), it is observed that the power consumptions of selfish nodes 3, 5 are much smaller than the fully generous cases (as shown by the arrows), and their total amount of power savings increases when selfishness index γ changes from 0.5 to 0.1. Meanwhile, for the generous nodes 1, 2, and 4, their power consumptions would have to increase to complement the selfish behaviors of nodes 3 and 5.

Figure 10.2(b) shows the total power consumption with different numbers of selfish nodes assigned in the aforementioned partial selfishness case. In contrast to the fully generous setting, partial selfishness leads to more energy consumption, and this gap becomes even larger when γ reduces significantly from 1 to 0.1, which means although selfish nodes save their own energy, the total energy consumption of five nodes as a group is higher. However, the all energy-aware selfish setting consumes the least energy in total. This is because by the rational energy-aware selfishness, power consumption is well balanced and no one is excessively depleted.

To summarize, the simulation results of the five-node example firmly demonstrate that our proposed approach captures the selfish behavior in an appropriate way, and the all energy-aware selfish setting achieves the best power savings, thus paving the way for its superior performance under a more practical and complex environment.

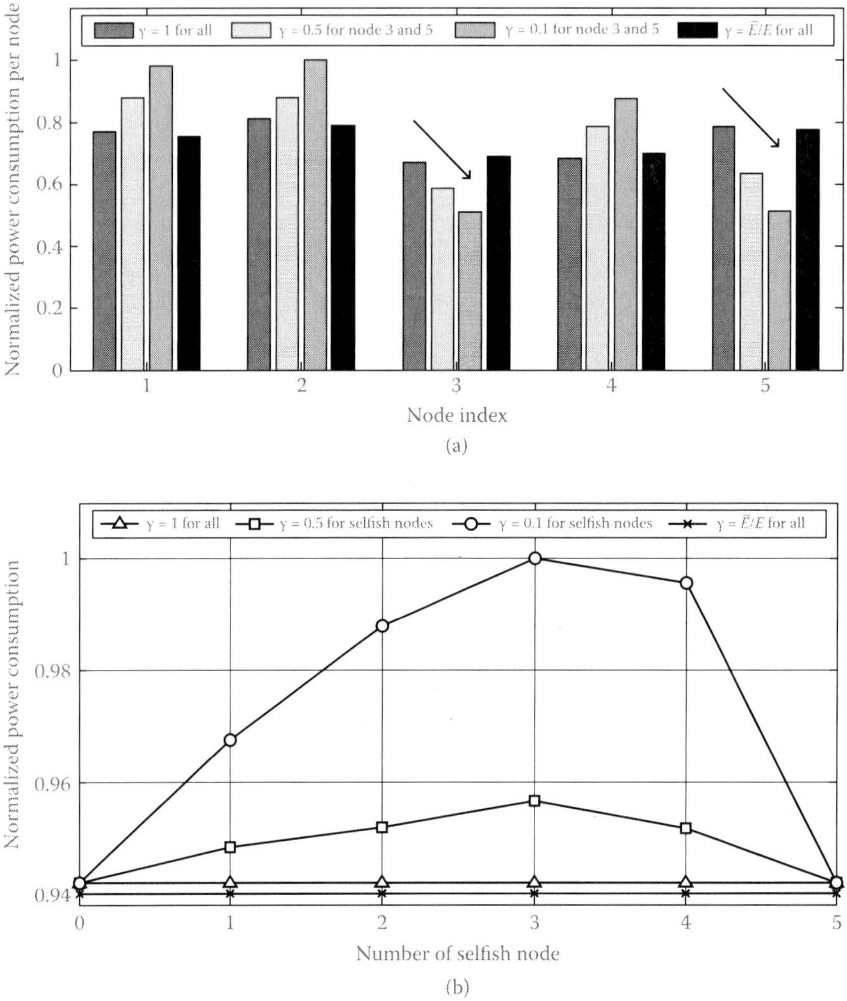

FIGURE 10.2 Simulation results for selfishness impact in four different conditions: fully generous case, partial $\gamma = 0.5$ selfishness, partial $\gamma = 0.1$ selfishness, and all energy-aware selfish case. (a) Normalized power consumption per node. (b) Total power consumption versus the number of selfish nodes.

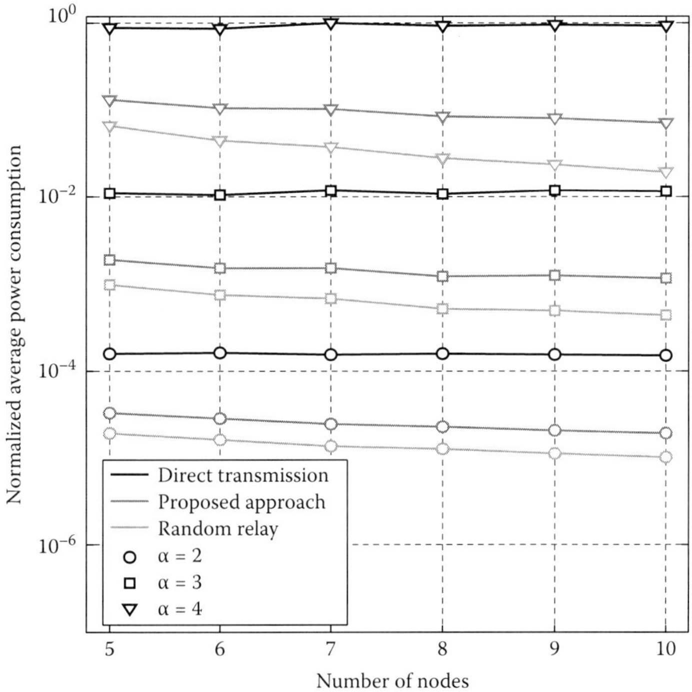

FIGURE 10.3 Normalized average power consumption per node versus the number of nodes, where $\alpha = \{2, 3, 4\}$.

10.5.2 A Complete Setting

In this scenario, we place N (varied between 5 and 10) nodes at uniformly random locations in a 100×100 m^2 region. A total of $1000 \times N$ packets are transmitted, and at each time t, the source and destination are randomly chosen. The simulation result is averaged over 100 runs for each N. We use the all energy-aware selfish setting and the weight function is also chosen as $g(\dot{U}) = e^{-\dot{U}}$. We aim to investigate the average power consumption per node with our proposed approach under different parameter settings (i.e., by varying the path-loss exponent, outage probability threshold, and desired data rate) and compare its performance with direct transmission and random relay approach. The random relay selection applies the same power allocation scheme with our proposed approach but selects the relay set randomly.

Figure 10.3 shows the average normalized power consumption per node with regard to different network sizes and α. We observe that our proposed

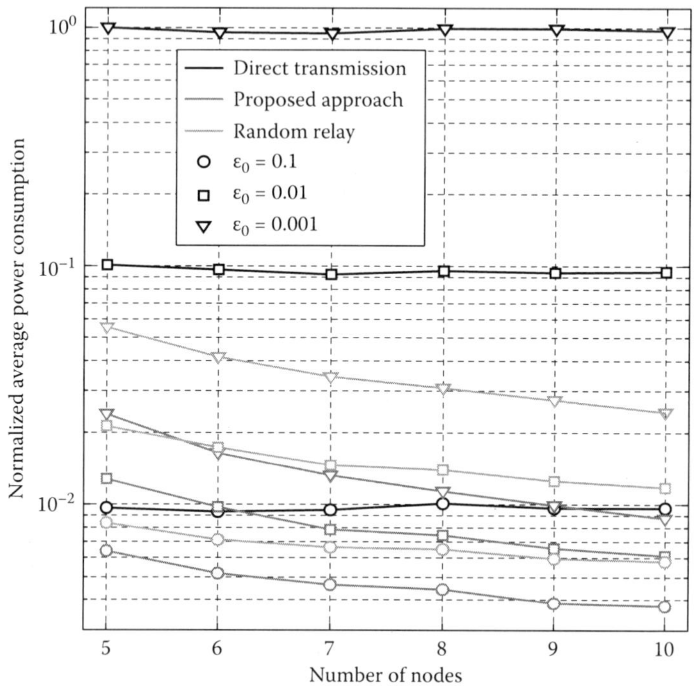

FIGURE 10.4 Normalized average power consumption per node versus the number of nodes, where $\alpha = 3$, $\epsilon_0 = \{0.1, 0.01, 0.001\}$, $R = 1$.

approach consumes less energy than both the random relay selection approach and direct transmission by a factor of 1/2 and 1/9 when $N = 5$, respectively; this gain becomes larger when N increases. Furthermore, as the network density increases, the average energy consumption of both the proposed approach and the random relay selection approach decreases significantly since nodes in the nearby proximity can be leveraged as relays. With the increase of α, it is not surprising that the network would spend more energy to combat the large-scale fading factor. The effects of different ϵ_0 and R can be seen in Figures 10.4 and 10.5, where we can observe the similar trends that a lower outage probability threshold and a higher data rate (more stringent QoS requirements) eventually lead to higher power consumption; however, the proposed approach always outperforms the other two schemes.

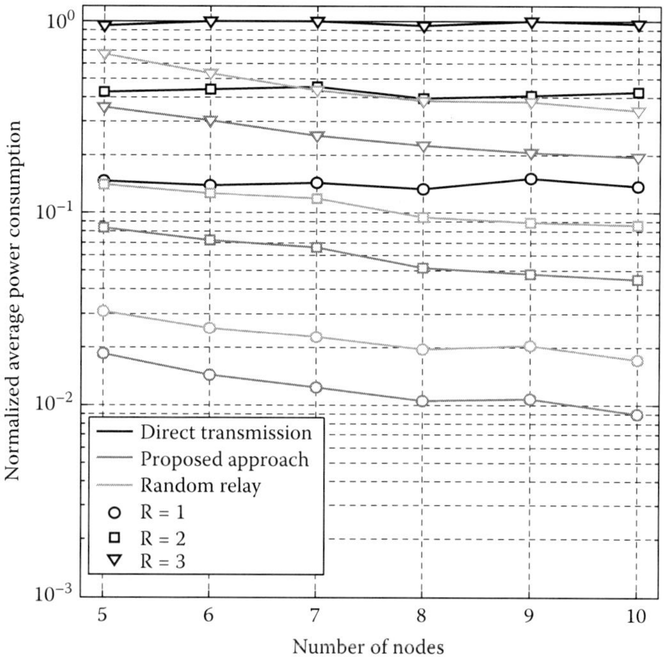

FIGURE 10.5 Normalized average power consumption per node versus the number of nodes, where $\alpha = 3$, $\epsilon_0 = 0.01$, $R = \{1, 2, 3\}$bps/Hz.

Table 10.1 illustrates the fairness performance by applying Jain's fairness index[3] to the relay's power in the proposed approach for the network parameter $\epsilon_0 = 0.01, R = 1$. It can be seen that the power allocation between multiple relays reaches a relatively high level of fairness and does not change dramatically over different large-scale fading factor α. This is achieved by the proposed utility function and the two-step relay selection algorithm, which balances the energy consumption of each node at each packet transmission.

10.6 CONCLUSIONS AND FUTURE WORK

In this chapter, we investigate the problem of relay selection in cooperative communication, focusing on multiple simultaneous relays. We first derive a closed-form expression of the outage probability under repetition-coded DF cooperation. Then, we introduce the novel concept of selfishness index to

[3]It is defined by $(\sum P_i^C)^2/(N\sum(P_i^C)^2), \forall i \in \mathcal{K}_s$. The result ranges from $\frac{1}{N}$ (worst case) to 1 (best case). The larger the index is, the better fairness that we can achieve.

TABLE 10.1 Fairness Index (Jain's) among the Relay Power

	$\alpha = 2$	$\alpha = 3$	$\alpha = 4$
$N = 5$	0.9484	0.9485	0.9457
$N = 8$	0.9478	0.9421	0.9465
$N = 10$	0.9443	0.9450	0.9473

capture the selfish behavior in cooperative transmission and incorporate it into a novel utility parameter representing the attained net payoff. Finally, we propose a two-step relay selection mechanism covering all aspects of power efficiency, energy fairness, and network lifetime, by using the utility function that makes the process reasonable and rational. Extensive simulation results show that our scheme outperforms both the direction transmission and random relay selection scheme in power savings and successfully achieves a satisfactory level of energy consumption fairness. In the future, we plan to investigate the buffer-aided relay selection and power allocation problem.

CHAPTER 11

Network Protocol Design of M2M-Based Cooperative Relaying

Zhengguo Sheng, Hao Wang, and Daqing Gu, Orange Labs Beijing, China
Xuesong Chen and Changchuan Yin, Beijing University of Post and Telecommunications, China
Chi Harold Liu, Beijing Institute of Technology, China

11.1 INTRODUCTION

Future wireless networks are expected to support a mixture of real-time applications, such as voice and multimedia streams [85], and non–real-time data applications, such as web browsing, messaging, and file transfers. Compared with wired environments, the associated communication channels and traffic patterns in mobile wireless networks are more unpredictable. Hence all of these applications impose stringent and diversified quality-of-service (QoS) requirements, which cannot be satisfactorily addressed through the traditional communication system. Recently, the availability of low-cost and high-processing-capability hardware that is capable of delivering multimedia content from the environment has fostered the development of wireless multimedia networks [86, 87, 88], that is, networks of resource-constrained wireless devices that can retrieve and deliver multimedia content such as voice and video streams, still images, messaging, and file transfers. As a result, it is predicted that wireless multimedia networks should require energy efficiency and reliable transmission links while keeping satisfactory QoS.

Toward this goal, multiple-input multiple-output (MIMO) has received significant attention; MIMO can provide spatial diversity and hence represents a powerful technique for interference mitigation and reduction [89, 90]. Although MIMO systems can show their huge benefit in cellular base scenarios, they may face challenges when it comes to their deployment in mobile devices. In particular, the typically small size of wireless devices makes it impractical to deploy multiple antennas. To overcome this drawback, the concept of cooperative communication mechanisms has been proposed as an effective way of exploiting spatial diversity to improve the quality of wireless links [2, 1, 91, 15, 92]. The key idea is to have multiple wireless devices in different locations cooperatively share their antenna resources and aid each other's wireless transmission effectively to form virtual and distributed antenna arrays. The previous work in the literature shows that cooperative communication can significantly improve the overall quality of the wireless transmission, in terms of the reception reliability [93, 1, 9], throughput [10], and power consumption [11].

In cooperative communication, the term *cooperation* refers to a node's willingness to share its own resources (e.g., energy, transmission opportunity) for the benefit of other nodes. It is thus important to understand how many resources must be consumed to reap the benefits of the cooperative communication. In our previous work [78], we have shown that it is advantageous to employ cooperative transmission in a network with multiple, mutually cooperative nodes, which can significantly reduce the total power consumption while maintaining a given level of QoS. However, there is no clear answer about whether cooperative communication requires more (or fewer) overall resources than conventional, non-cooperative communication to achieve the same level of wireless link quality. If so, by how much can we best achieve the resource savings when employing cooperative communication? What is the impact of each node's "willingness" to cooperate on energy efficiency when selfish and unselfish natures are imposed on individuals? What are the applicable scenarios and how can we incorporate the proposed solution into distributed MAC and routing protocols? This chapter is our answers to these questions, with particular focus on the energy consumption issues in cooperative communication to support multimedia services with stringent QoS requirements.

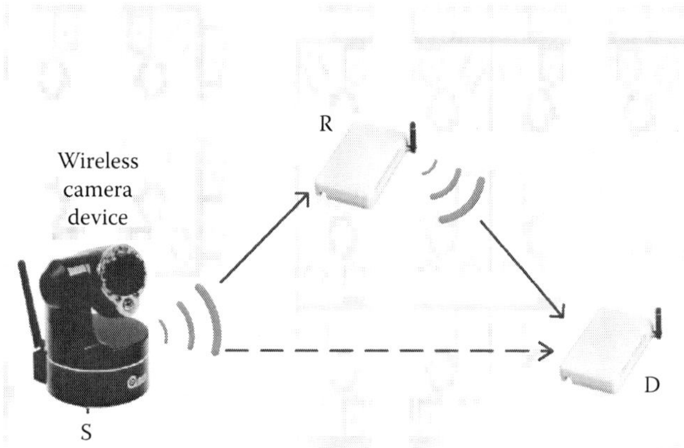

FIGURE 11.1 An example of a wireless cooperative link.

We consider in this chapter the decode-and-forward (DAF) cooperative communication. Figure 11.1 illustrates the concepts of DAF, where the transmission of a source (S) to a destination (D) is aided by a relay (R). While there are variants of cooperative communication, depending on how the relay cooperates to the source's transmission, R's action in DAF is to overhear the packet transmitted by S, decode it, and retransmit it to D, improving the reception quality of the (combined) signal at D. This cooperative scheme lends itself to a relatively easy implementation in hardware and software. Moreover, it is shown in Laneman et al. [1] that such a cooperative scheme can achieve full second-order diversity and therefore provides significant improvement to reception reliability.

Specifically, we explore the energy consumption aspect of DAF cooperative communications (CC) from various angles. First, we look at how much transmitting power is required for the source and the relay in the cooperative transmission for a given requirement on the link quality. This result is then used to investigate how much and in which case the power can be saved by using cooperative communication, compared to conventional and (non-cooperative) direct communication. Based on these analytical results, we propose a strategy for a resource allocation problem in networks of multiple cooperative nodes, namely the energy-efficient relay-selection rule for each packet transmission. In order to study the achievable energy savings due to cooperative communications at a fundamental level, we assume throughout the chapter that a predefined QoS requirement in terms of the transmission rate and the outage probability is given.

The following summarizes our contributions and key results:

- We derive a closed-form solution for the optimal transmission power required by each source and relay node in DAF cooperative communication under a Rayleigh fading channel model to achieve the given QoS requirements. Under the optimal power allocation, our analysis shows that the required transmission power of the relay is always smaller than that of the source, a result that lays a foundation to encourage the cooperative behaviors as this means that the helping party (relay) only needs to spend a relatively small amount of energy compared with the one seeking help from others (source).
- We analyze the power consumption in the optimal cooperation scheme and compare its performance with both direct transmission and conventional cooperation where both source and relay power are considered to be identical. Specifically, we define the term of power efficiency and investigate the best relay location to achieve the minimum power consumption as well as derive the bound performance compared with conventional cooperation.
- We propose adaptive relay selection rule that will help select appropriate relays for the fairness and maximal energy savings of each node in a multi-node environment and analyze each node's "willingness" to cooperate when selfish and unselfish natures are imposed to individuals as well as incorporate them into the implementation of distributed medium access control (MAC) and machine-to-machine (M2M) routing protocols. Simulation results are supplemented to illustrate the significant energy savings of the proposed relay selection rules in providing reliable services.

This chapter is organized as follows. The analysis of the optimal cooperation scheme is presented in Section 11.2. The energy-efficient relay-selection rules and their practice in MAC and routing protocols are proposed in Section 11.3 and simulation results are provided in Section 11.4. Finally, concluding remarks are given in Section 11.5.

11.2 ANALYSIS OF OPTIMAL DAF COOPERATION

In this chapter, we consider the same system model as in Chapter 7, and we provide a comprehensive analysis of the optimal DAF cooperative transmission. First, we discuss the best relay location for optimal DAF, which can achieve the maximum energy savings compared with direct transmission. Then, we analyze its advantage in energy savings, by comparing it with a conventional cooperative scheme.

11.2.1 Power Efficiency Factor

We introduce a power efficiency factor β that represents the ratio of the total transmission power of cooperative transmission to that of direct transmission:

$$\beta = \frac{p+q}{p_D}. \tag{11.1}$$

Clearly, small values of β are always preferable.

According to (2.6) and (7.6), the power efficiency factor for optimal DAF is defined by

$$\beta^* = \frac{p^*+q^*}{p_D} = \frac{\sqrt{\frac{m+1}{4}}\left(\sqrt{d_{s,r}^{\alpha}} + \frac{d_{r,d}^{\alpha}}{m\sqrt{d_{s,r}^{\alpha}}}\right)}{K\sqrt{d_{s,d}^{\alpha}}}, \tag{11.2}$$

where $K = ((2^b+1)\sqrt{2\epsilon^{\text{out}}})^{-1}$ is the QoS factor, $\gamma = d_{r,d}^{\alpha}/d_{s,r}^{\alpha}$, and $m = (\gamma + \sqrt{\gamma^2+8\gamma})/2$.

11.2.2 Best Relay Location for Optimal DAF

Result 11.1 *For any relay r that is non-collinear with the source s and destination d, we can always find a mapping relay r' on $\overline{sd}$ that achieves a lower total power consumption, given the same target QoS.*

Proof As can be seen from Figure 11.2, given a relay r that is outside the line $\overline{sd}$, we can find a point r' on $\overline{sd}$ as the mapping relay where $\overline{rr'}$ is perpendicular to $\overline{sd}$. Clearly, we have $d'_{s,r} < d_{s,r}$, $d'_{r,d} < d_{r,d}$, and hence, $a' < a$, $b' < b$.

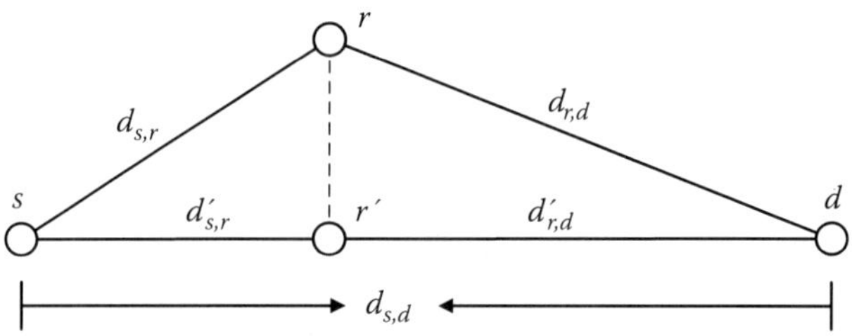

FIGURE 11.2 The position of a mapping relay.

From (7.6), we define $f(a,b) = p^* + q^*$. Since $\frac{\partial f(a,b)}{\partial a} > 0$ and $\frac{\partial f(a,b)}{\partial b} > 0$, we can obtain $p^* + q^* = f(a,b) > f(a',b) > f(a',b') = p'^* + q'^*$, which completes the proof. □

Theorem 11.1 *For path loss $\alpha = 2$, the best relay location that minimizes β^* for the optimal DAF cooperation is at the destination.*

Proof From Result 11.1, we can find that the relay location that minimizes β^* for the optimal DAF cooperation is surely on the line $\overline{sd}$, namely $d_{s,r} + d_{r,d} = d_{s,d}$. Bringing this result into (11.2), we can obtain the ratio for $\alpha = 2$:

$$\beta^* = \frac{1}{K}\sqrt{\frac{m+1}{4}}\left(1 + \left(\sqrt{\frac{\gamma}{m^2}} - 1\right)\frac{d_{r,d}}{d_{s,d}}\right),$$

where $\gamma d^{\alpha}_{r,d}/d^{\alpha}_{s,r}, m = (\gamma + \sqrt{\gamma^2 + 8\gamma})/2$. Since $\beta^* \geq 0$, it is easy to observe that the minimum value can be obtained as $\frac{1}{2K}$ when $\frac{d_{r,d}}{d_{s,d}} = 0$. □

Note that this result is different from that of the conventional cooperative scheme with identical power [94], where for $\alpha > 1$ the best relay location for DAF cooperation is proved to be the midway between source and destination.

11.2.3 Comparison with Existing Literature

Most recent literature on single relay selections is based on the identical power assumption for both source and relay (for example, [94, 14, 95, 16]). To better evaluate the performance of the optimal DAF cooperation and compare its performance with existing solutions, we consider this equal power scenario as the conventional cooperative scheme where the source and relay nodes always use the identical transmission power, that is, $p = q = p_{\text{con}}$. Hence the minimum total power consumption can be derived from (7.4) as follows:

$$2p_{\text{con}} = 2\sqrt{\frac{1}{2}d_{s,d}^{\alpha}(d_{s,r}^{\alpha} + d_{r,d}^{\alpha})\frac{(2^{2b}-1)^2}{\epsilon_C^{\text{out}}}}. \tag{11.3}$$

Then, the power efficiency factor is defined by

$$\beta' = \frac{2p_{\text{con}}}{p_D} = \frac{\sqrt{d_{s,r}^{\alpha} + d_{r,d}^{\alpha}}(2^b+1)\sqrt{2\epsilon^{\text{out}}}}{\sqrt{d_{s,d}^{\alpha}}}. \tag{11.4}$$

Theorem 11.2 *Given a relay r, the power efficiency of optimal DAF* (11.2) *is always lower than that of conventional cooperative scheme* (11.4)*, and we have the bound performance of optimal DAF as*

$$\frac{\beta'}{2} < \beta^* < \beta'. \tag{11.5}$$

1. *The lower bound is obtained when the relay approaches the destination, where we have the optimal source power* $p^* = p_{\text{con}}$ *and the performance gain can reach to its maximum with the relay power* q^* *down to 0.*

2. *The upper bound is obtained when the relay approaches the source node, where we derive* $p^* = q^* = p_{\text{con}}$*, which means that the optimal cooperation uses the same amount of power as the conventional cooperation.*

3. *When the relay goes to infinite or stays in the middle of source and destination, we have* $\frac{\beta^*}{\beta'} = \sqrt{\frac{27}{32}}$.

Proof From (7.6), we can obtain the power ratio of the optimal source power to the optimal relay power as follows:

$$\frac{p^*}{q^*} = \frac{1}{2}\left(1 + \sqrt{1 + \frac{8}{\gamma}}\right) > 1. \tag{11.6}$$

From (7.6) and (11.3), we have the power ratio of the optimal source power to the conventional source power as follows:

$$\begin{aligned}\frac{p^*}{p_{\text{con}}} &= \sqrt{\frac{d_{r,d}^{\alpha} + \sqrt{d_{r,d}^{2\alpha} + 8d_{r,d}^{\alpha}d_{s,r}^{\alpha}} + 2d_{s,r}^{\alpha}}{2(d_{s,r}^{\alpha} + d_{r,d}^{\alpha})}} \\ &= \sqrt{\frac{1}{2}\left(1 + \frac{1+\sqrt{\gamma^2+8\gamma}}{1+\gamma}\right)}, \end{aligned} \tag{11.7}$$

where $\gamma = \frac{d_{r,d}^{\alpha}}{d_{s,r}^{\alpha}}$. Bringing (11.6) into (11.7), we can derive

$$\frac{\beta^*}{\beta'} = \sqrt{1 + \frac{1+\sqrt{\gamma^2+8\gamma}}{1+\gamma}}\left(\frac{\sqrt{2}}{4} + \frac{\sqrt{2}}{2+2\sqrt{1+\frac{8}{\gamma}}}\right). \tag{11.8}$$

Note that (11.8) is a function of γ. Then, we have

$$\frac{\beta^*}{\beta'} = g(\gamma),\ \forall \gamma > 0. \tag{11.9}$$

Since $\frac{\mathrm{d}g(\gamma)}{\mathrm{d}\gamma} > 0$, we can get the bound of $g(\gamma)$ as

$$\begin{aligned} g_{\min}(\gamma) &= \lim_{\gamma\to 0} g(\gamma) = \frac{1}{2}, \\ g_{\max}(\gamma) &= \lim_{\gamma\to\infty} g(\gamma) = 1. \end{aligned} \tag{11.10}$$

Therefore, we have the result in (11.5). Additional comments to the bound performance are explained as follows:

1. Lower bound: This is obtained by putting $\gamma = 0$ in (11.6) and (11.7). It is worth noting that the optimal relay power highly depends on the relay location and has

$$q^* \to 0, \quad \text{if } d_{r,d} \to 0. \tag{11.11}$$

   In that case, the whole receiver side can actually be treated as a MIMO antenna system with the relay and destination combined together.

2. Upper bound: This is obtained by putting $\gamma = \infty$ in (11.6) and (11.7).
3. The result $\frac{\beta^*}{\beta'} = \sqrt{\frac{27}{32}}$ is obtained by putting $\gamma = 1$ in (11.6) and (11.7). □

Theorem 11.3 *Compared with the transmission power of the conventional cooperative scheme, the optimal source power* p^* *is bounded by*

$$p_{\text{con}} < p^* < 1.23 p_{\text{con}}. \tag{11.12}$$

Proof We first derive the upper bound of $\frac{p^*}{p_{\text{con}}}$. It is observed that if (11.7) can have the maximum value, then its inner term $(1 + \sqrt{\gamma^2 + 8\gamma})/(1+\gamma)$ should be maximum. Note that $h(\gamma) = (1 + \sqrt{\gamma^2 + 8\gamma})/(1+\gamma)$ is a convex function for $\gamma > 0$ since $\frac{\mathrm{d}^2\, h(\gamma)}{\mathrm{d}^2\,\gamma} \leq 0$. Hence there will be only one maximum for $\gamma > 0$. Taking the first order of $h(\gamma)$, we have the optimal γ^* to get the maximum $h(\gamma)$ as $\gamma^* = 2 - \sqrt{2}$. Replacing it in (11.7), we can obtain the upper bound.

From (11.7), the lower bound performance is achieved when the relay node approaches the source or the relay node approaches the destination, that is, $\gamma = 0$ or $\gamma = \infty$. □

In general, Theorem 11.3 tells us that the optimal source actually spends more power than that in the conventional cooperation. However, from (11.6) and Theorem 11.2, a careful reader might notice that the optimal cooperation can help significantly reduce the relay power, especially when the relay approaches the destination. In other words, slightly increasing the source power can help significantly reduce the relay power and thereafter save the total energy.

11.3 COOPERATION-AIDED ROUTING IN LOW-POWER AND LOSSY NETWORKS

In low-power and lossy networks (LLNs), mobile devices typically operate with constrained memory, processing power, and energy, and their interconnections are typically characterized as unreliable links with high loss rates. RPL [96] is a routing protocol designed for LLNs, which is a de facto M2M standard to support multimedia services provided by upper layer protocol, that is, Constrained Application Protocol (CoAP). RPL is a distance vector routing protocol, in which nodes construct a destination-oriented acyclic graph (DODAG) by exchanging distance vectors and root to a controller. Through broadcasting routing constraints, the root node (i.e., central control point) filters out the nodes that do not meet the constraints and selects the optimum path according to the metrics (e.g., hop count, energy cost, latency). In the stable state, each node has identified a stable set of parents and forwarded packets along the path toward the "root" of the DODAG.

However, the current solution cannot well support multimedia services in wireless networks; for example, transmitting images in a multi-hop fashion in a harsh outdoor environment to monitor emergency accidents may result in high packet loss. Moreover, because of the hierarchical transmission structure, the nodes approaching the root may experience more traffic and energy consumption, thus being vulnerable to energy depletion. To address these issues and improve the reliability of wireless routing in low-power and lossy networks, we incorporate CC into the RPL protocol and propose the cooperation-aided routing protocol for lossy networks. Specifically, in the stage of topology formation, each node should maintain two tables: the routing table, a list of parents toward the root; and the relay table, a set of candidate nodes that can serve as the relay between the node itself and its parents. Each node builds up its routing table through DODAG information object (DIO) messages. Neighboring nodes periodically exchange routing tables to check whether they have the same parent. If so, each of them will be selected as a candidate relay for the other and added to the corresponding relay table. In this way, the relay table constructs a relay link between both sides where cooperative transmission can be performed. An example of this process is shown in Figure 11.3, where nodes B and C serve as the candidate relay for each other because they share a common parent node A.

Finally, when a node transmits its packet toward the root, the next hop is determined by the routing table. If its relay table is not empty, the node itself will select one relay from the candidates according to the weighted adaptive relay selection rule in Section 7.2.1 and perform the optimal DAF cooperative transmission. Therefore, enhanced reception reliability and reduced energy consumption are expected during the transmission of each hop.

11.4 PERFORMANCE OF COOPERATION-AIDED ROUTING

In this scenario, we consider a grid network topology for multipoint-to-point simulation, where the designated root is located at the center of a 100 m ×

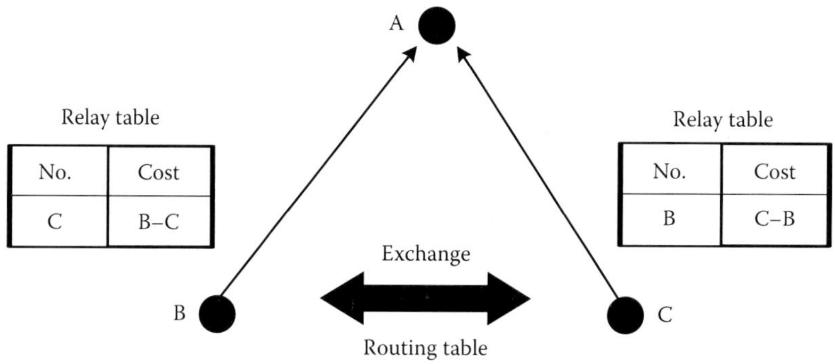

FIGURE 11.3 An example of building up relay tables.

100 m region, with 12 surrounding nodes in the peripheral area. As shown in Figure 11.4(a), according to the number of hops to the root, the surrounding nodes can be divided into two categories based on the RPL routing protocol, namely the rank 1 nodes 2–5 and the rank 2 nodes 6–13. Note that the solid lines are next-hop links and the dashed lines are relay links, as discussed in Section 11.3. Throughout the simulation, we set the path-loss exponent $\alpha = 3$, the data rate $b = 1$ bps/Hz, and the targeted $\epsilon^{\text{out}} = 0.01$. The initial payoff value of every node ($u_i(0)$) is set to 0. The transmission range is 45 m. A total of 1000 packets are transmitted to the root from random selected sources using RPL routing.

Figure 11.4(b) shows the normalized total energy consumption of each node. Obviously, each node consumes less energy in the weighted adaptive relay approach compared to the direction transmission. Furthermore, it is worth noting that the performance of lower-rank nodes (close to the root) with cooperation is closed to the performance of higher-rank nodes without cooperation, which shows that the lower-rank nodes with heavy traffic actually benefit more from cooperative transmission. Moreover, we can observe that the proposed relay selection scheme can fairly distribute the energy consumption among nodes with the same rank. This is so because our proposed scheme prioritizes the selection of relays with larger payoff value, while in turn the relaying transmission reduces its cumulative payoff, thus balancing the opportunity of being selected among all nodes.

As a different example, we consider a random network topology as shown in Figure 11.4(c), where the root stays at (40, 40) with nine randomly located nodes. Without changing the parameters, the routing path is generated by RPL with next-hop links (solid lines) and relay links (dashed lines).

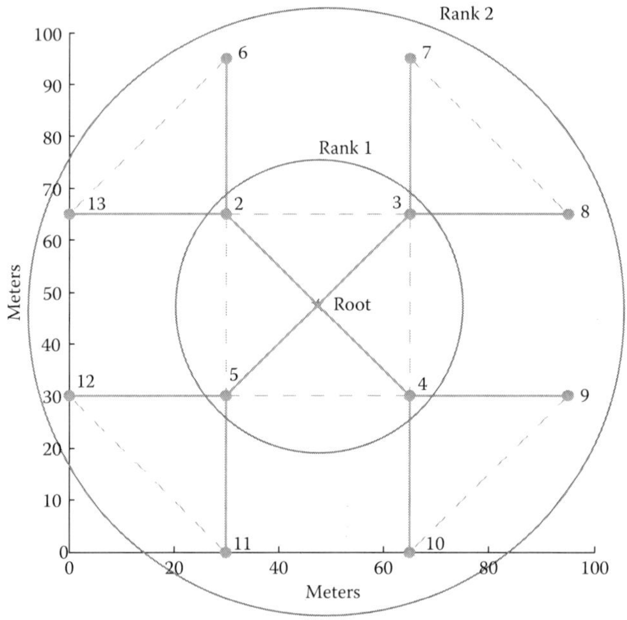

(a) A grid network scenario.

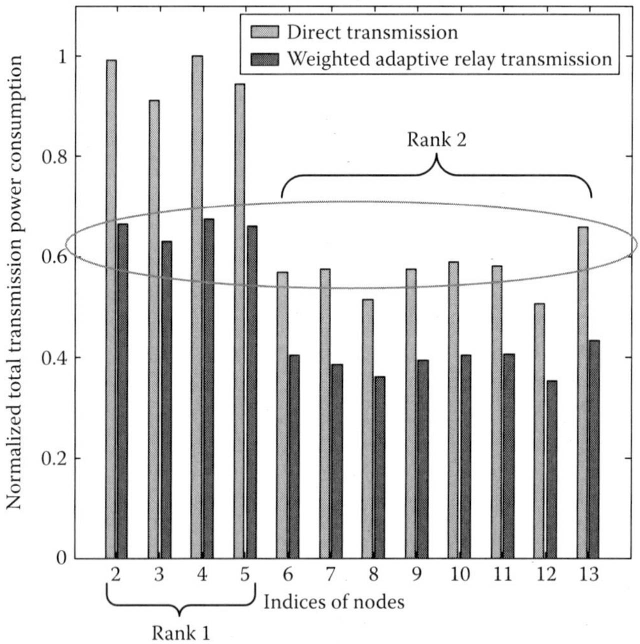

(b) Normalized total transmission power consumption in the grid network scenario.

FIGURE 11.4 Simulation results.

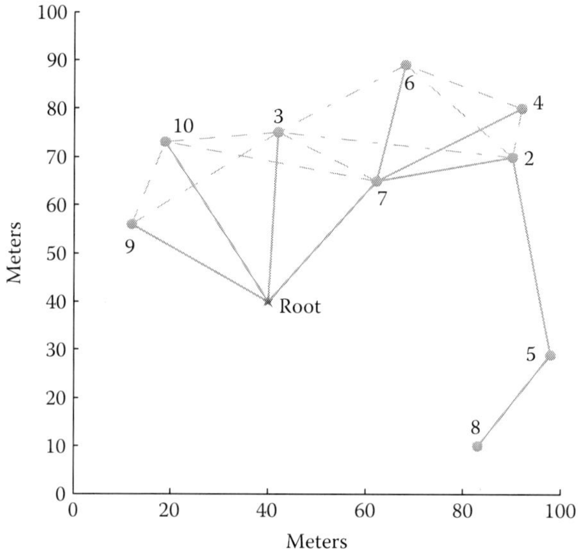

(c) A random network scenario.

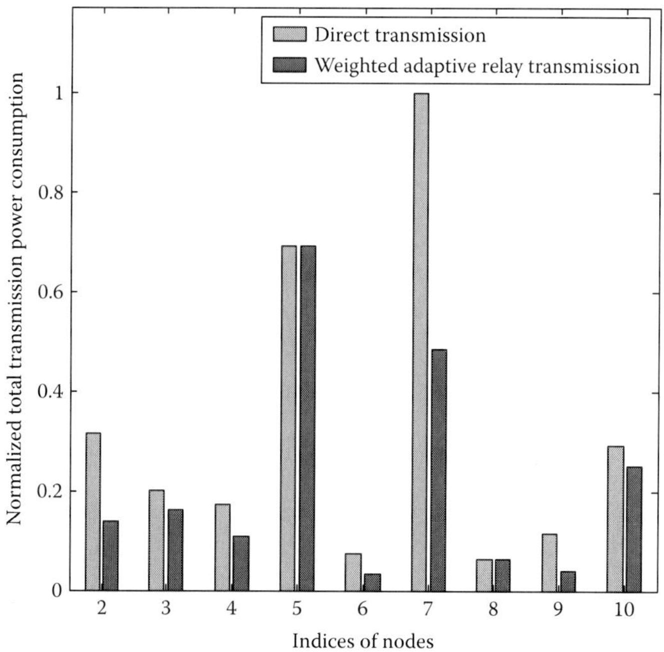

(d) Normalized total transmission power consumption in the random network scenario

FIGURE 11.4 *(continued)* Simulation results.

To analyze the simulation result, we divide the nine nodes into two groups: the cooperation group (nodes 2, 3, 4, 6, 7, 9, 10) and direct transmission group (nodes 5, 8).

Figure 11.4(d) shows the comparison of normalized total energy consumption of cooperation-aided routing with that of using direction transmission. Overall, the nodes in cooperation group can significantly reduce their energy consumption because of the help from relaying; for nodes 5 and 8, because there are no potential relays nearby, the energy consumption is the same with direct transmission. Furthermore, it is worth noting that the nodes especially with heavy traffic load (e.g., nodes 2 and 7) successfully gain a satisfactory level of benefit from cooperation, which is consistent with our results in Figure 11.4(b).

11.5 CONCLUSIONS

We have shown in this chapter that it is advantageous to allocate non-uniform powers to various cooperative transmitters in wireless multimedia networks, which can significantly reduce the total power consumption while maintaining a given level of quality of service (QoS). Specifically, we have proposed an optimal power-allocation method for the decode-and-forward (DAF) wireless cooperative networks and investigated its power efficiency. Our analysis shows how the proposed DAF cooperation outperforms the conventional cooperation and direct transmission. We have also introduced the adaptive relay-selection rule that can serve as an effective tool to achieve a desirable trade-off between fairness and energy consumption at each node, and demonstrated the advantages in MAC and M2M routing protocols.

To develop further robust cooperative schemes to cope with new demands in future wireless multimedia networks, we plan to explore the performance gain of the cooperative relay-selection methods and propose additional robust relay-selection mechanisms. For example, the additional mechanisms must be able to consider network scenarios where nodes can have different traffic loads, while maintaining a satisfactory degree of fairness. We also plan to consider the lifetime or the remaining energy of each node as input parameters to the fairness measure.

CHAPTER 12

Conclusion

Zhengguo Sheng, University of British Columbia, Canada
Chi Harold Liu, Beijing Institute of Technology, China

12.1 CONTRIBUTIONS AND CONCLUSIONS

The main focus of this book has been the development of performance effective algorithms for cooperative wireless networks. The performance of cooperative communication is measured based on various criteria, such as cooperative region, average power ratio, ETE reliability (outage), ETE energy consumption, ETE throughput, and ETE delay. In the following sections, we briefly highlight the important contributions and conclusions of this work.

12.1.1 Fundamental Understanding of Cooperative Routing

The main contribution of Part I is to analyze and compare the performance of end-to-end reliability, energy consumption, throughput, and delay of wireless cooperative communication from the network layer's perspective. In essence, these benefits make cooperative wireless networks capable of combating radio unreliability and meeting future application requirements of high-speed and high-quality services with high energy efficiency. The acquired new insights on the network performance can also provide a precise guideline for the efficient designs of practical and reliable communications systems. The detailed contributions include the following:

- Cooperative routing algorithms
- Interference subtraction and supplementary cooperation

- End-to-end analysis of cooperative routing (i.e., reliability, energy, throughput, and delay)

12.1.2 Fundamental Understanding of Cooperative Communication Using Probabilistic Tools

The main contribution of Part II is to characterize the energy performance of cooperative transmission from a single link's perspective, which provides a fundamental understanding of cooperative transmission. Such understanding illustrates a better vision on how to design an efficient cooperative transmission strategy and utilize it on upper layers. In particular, we have defined the notion of a *cooperative region* and the *average power ratio* of DAF cooperative transmission. Opportunities for cooperation increase in harsher environments: as the source-destination distance $d_{s,d}$ or the path-loss exponent increase, or as the desired outage probability decreases. More specifically, the developed algorithms include the following:

- Cooperative region and average power ratio
- Optimal power allocation method for the DAF cooperative wireless networks
- Energy-efficient relay selection for cooperative wireless networks

12.1.3 Cooperative Communication in Practice

Part III has extended our focus from a cooperative route to a general cooperative network with multiple source-destination pairs and employed different cooperation strategies to improve wireless communication performance. The detailed contributions include the following:

- Energy-aware cooperative communication
- Joint beamforming and optimal power allocation for cooperative communication
- Network protocol design for cooperative relaying in M2M communication

12.2 FUTURE WORK

We now identify some areas that can extend the current state of the art in this book.

12.2.1 Robust Relay Selection Schemes

Considering the resemblance to the study on evolutionary games [97], we are keen to pursue a future research topic of in-depth analysis of the dynamics and the emergence of cooperation in cooperative communication, possibly using analytical tools such as population dynamics.[1]

To develop further robust cooperative schemes to cope with new demands in our future wireless networks, we plan to explore the performance gain of the cooperative relay-selection methods and also propose additional robust relay-selection mechanisms, which can account for other parameters of importance as the fairness measures, such as the difference in traffic loads and the remaining energy of each node.

12.2.2 A Cross-Layer Design for Joint Flow Control, Cooperative Routing, and Scheduling in Multi-hop Wireless Sensor Networks

Cooperative communication schemes open up a new dimension to design the upper-layer networking protocols. There have been several studies that explore the advantages by using cooperative transmission schemes at the physical layer to increase the performance of upper layers in wireless sensor networks (WSNs) (e.g., [98, 99]) and other types of wireless networks (e.g., [100, 22, 101]). However, all of them mainly focus on medium access control (MAC) and network layers, or cross-layer designs from physical to network layers. To the best of our knowledge, the transport layer issues on fairness of multiple end-to-end competing flows in WSNs (and other types of wireless networks) using cooperative communications have never been considered yet.

In this future work, we focus on applying cooperative diversity to the network utility maximization (NUM) based cross-layer flow control framework [102], which deals with congestion control and fair resource allocation

[1] In fact, the pairwise payoff structure of cooperative communication is that of the Prisoner's Dilemma, which is one of the most studied subjects in evolutionary game theory.

by regulating the generating rate of every source node to maximize the aggregate utility[2] function of source nodes at the transport layer, considering various constraints at lower layers. In particular, we consider how to use two basic cooperative schemes, broadcasting and beamforming [22, 101], to improve the global end-to-end throughput. To this end, the following four issues must be considered:

1. The physical layer capacity of the cooperative transmission links (i.e., the maximal allowed transmitting rate over a broadcasting or beamforming link with the given bit error requirement).

2. The interference among direct and cooperative links, which greatly limited the network throughput. In particular, we consider the time division multiple access (TDMA) approach to schedule the activity of different links (see, e.g., [103, 104]).

3. For WSNs with arbitrary topologies, how to choose the end-to-end forwarding routes of different flows, consisting of sequences of hybrid one-hop direct transmission and cooperative transmission decisions [101].

4. Adapting a source rate (sensing rate of every sensor node) that can achieve maximal aggregate network utility and guarantee all lower layer constraints.

12.2.3 Cooperative Communications in VANETs

Cooperative transmission has been shown to be a low-cost and spectrally efficient technique to combat small-scale multipath fading in wireless networks. However, reliable design of communication protocols for vehicular ad hoc networks (VANETs) is particularly challenging due to the fluctuating quality of radio channels and the constant changes of network topology. To resolve these issues, we first enhance radio links by distributed beamforming and relay-selection techniques using the knowledge of dynamic vehicular density in urban environments. We also plan to develop new cooperative transmission protocols for VANETs with short-lived radio links with fluctuating quality due to the high mobility of vehicles. Such protocols will include efficient mechanisms for identifying appropriate relay nodes despite mobility.

[2]A utility function is generally used to characterize the network performance such as the fairness of difference flows and the global end-to-end throughput.

Vehicles equipped with suitable devices are capable of accessing the Internet using roadside infrastructure. A natural extension is the combination of mobile nodes (vehicles) with stationary gateways installed along the road to boost performance. Cooperative communication between mobile nodes and gateways will be studied as an alternative, low-cost way to enhance reliability and connectivity for vehicle-to-vehicle communications. To support high data throughput, we plan to maximize the degree of concurrent transmissions for the developed cooperative protocols involving the gateways, while maintaining the reception in terms of reliability, error rate, and packet delay at a satisfactory level.

Appendix

A.1 OPTIMAL COOPERATIVE ROUTE

Typically, delay is strongly related to the number of hops in a route. In the context of cooperative networks, one hop can be a direct link or a cooperative link, as defined in Section 2.1. A meaningful routing problem is to find a route with no more than N hops that minimizes the outage performance in the cooperative networks. Based on the analysis of the optimal relay location problem in Section 2.3, the ϵ^{out} for the cooperative link can be minimized by locating the relay at the middle of the node pair associated with the link. For a route with multiple cooperative links, it is obvious that "straight line" routes can achieve better outage performance than any other curve-shaped routes. Furthermore, one can observe that the route that minimizes the ϵ^{out} must have the maximum allowable number of hops N. By assuming that the error performances among links are independent in a given cooperative network, the ETE outage probability is given by

$$\epsilon_{\text{ETE}} = 1 - \prod_{i \in N} (1 - \epsilon_i^{\text{out}}),$$

where ϵ_i^{out} denotes outage probability for the cooperative link i. For small outage probabilities $\epsilon_i^{\text{out}} \ll 1$, we make the following approximation

$$\epsilon_{\text{ETE}} \approx \sum_{i \in N} \epsilon_i^{\text{out}}.$$

Based on these observations, the routing optimization problem becomes

$$\begin{cases} \min \quad \epsilon_{\text{ETE}} = \displaystyle\sum_{n=1}^{N} d_{i,j}^{n\;2\alpha} \frac{(2^{2b}-1)^2}{2^{\alpha} p^2} \\ \text{s.t.} \quad \displaystyle\sum_{n=1}^{N} d_{i,j}^{n} = D, \end{cases} \tag{A.1}$$

where $d_{i,j}^n$ is the distance between node i and j associated with the nth link in the route, and D is the total distance along the route from the source to the destination. We can then simplify the problem and obtain the Lagrangian for this problem as

$$L = \sum_{n=1}^{N} d_{i,j}^{n\ 2\alpha} + \lambda \left(D - \sum_{n=1}^{N} d_{i,j}^n \right).$$

The conditions for optimality are

$$\frac{\partial L}{\partial d_{i,j}^n} = 2\alpha d_{i,j}^{n\ 2\alpha-1} - \lambda = 0\,.$$

Hence $d_{i,j}^n = \sqrt[2\alpha-1]{\lambda/2\alpha}$. Substituting the results into (A.1) yields $d_{i,j}^n = D/N$. Clearly, in order to achieve the best outage performance, the cooperative links of the optimal routing are uniformly distributed along the line between the source and the destination node.

A.2 PROOF OF THEOREM 2.2

In order to achieve the minimal ETE outage probability in such a regular linear topology, the optimal solution for cooperative routing is shown below.

For even n: There is an odd number of links. Hence, the outage probability of optimal routing according to Section A.1 that can achieve the minimal ETE outage probability from the source to the destination is

$$\epsilon_{\text{optimal}} = \frac{(2^{2b}-1)^2}{2p^2} D^{2\alpha}(2^{\alpha}n - 2\alpha + 2 + 6^{\alpha} + 3^{\alpha})\,.$$

For odd n: There is an even number of links. Hence, all the cooperative links can be equally distributed. Then, the corresponding minimal ETE outage probability from the source to the destination is

$$\epsilon_{\text{optimal}} = \frac{(2^{2b}-1)^2 2^{\alpha-1} D^{2\alpha}(n-1)}{p^2}\,.$$

The proposed cooperative route is slightly different from the optimal solution and is more complicated to analyze. Using our proposed routing algorithm, we obtain the minimal ETE ϵ^{out} as follows:

1. If $\log_2(n)$ = integer or $\log_2(n-1)$ = integer, there is no difference between the route generated by the proposed algorithm and the optimal solution.
2. If $\log_2(\frac{n-1}{3})$ = integer, the gap ratio is $g = \frac{11}{4}$.

For any value n which satisfies the above condition, we then can obtain the ETE

$$\epsilon_{\text{proposed}} = \frac{(2^{2b}-1)^2}{2p^2} D^{2\alpha}(n2^{\alpha} - 2^{\alpha}13 + 3^{\alpha}4 + 2^{\alpha+2}3^{\alpha})$$

by using the proposed algorithm. Placing this into (2.19) yields $g = \frac{11}{4}$. However, compared with the optimal solution, we can reduce $2^{\log_2 \frac{n-1}{3} - 1}$ hops and $\frac{n-1}{3}$ nodes involved.

Otherwise, the gap ratio for odd number nodes is $\frac{33}{2(n-1)}$.

This proof is similar as above. Using the same argument, the ETE outage probability is

$$\epsilon_{\text{proposed}} = \frac{(2^{2b}-1)^2}{2p^2} D^{2\alpha}(n2^{\alpha-1} - 2^{\alpha-1}7 + 3^{\alpha} + 6^{\alpha})$$

and the gap ratio is $\frac{33}{2(n-1)}$. However, compared with the optimal solution, we can reduce 1 hop and 2 nodes involved.

A.3 DERIVATION OF (5.18) AND (5.19)

Define $x_n = \epsilon_{i,j}^{\text{DT}}$ and

$$K_n = \frac{1}{\rho \ln\left(1 + \frac{\text{SINR}_{i,j}^{n}}{1+\rho}\right)},$$

where $n \in [1, \cdots, |\mathcal{S}_1|]$, the first optimization problem in (5.17) can be written as

$$\begin{aligned} \min \quad & \sum_{n=1}^{|\mathcal{S}_1|} K_n(\rho L - \ln x_n) \\ \text{s.t.} \quad & \sum_{n=1}^{|\mathcal{S}_1|} x_n \le z. \end{aligned}$$

According to the Kuhn-Tucker condition, the inequality constraints can be converted to the equality constrains, and the optimal solution of x_n can be found from Boyd and Vandenberghe [69]:

$$\begin{aligned} -\frac{K_1}{x_1} + \lambda &= 0\,, \\ &\vdots \\ -\frac{K_{|\mathcal{S}_1|}}{x_{|\mathcal{S}_1|}} + \lambda &= 0\,, \\ \lambda\left(\sum_{n=1}^{|\mathcal{S}_1|} x_n - z\right) &= 0\,. \end{aligned}$$

Hence the Lagrange multiple and the optimal solutions of x_n should be

$$\lambda = \frac{1}{z}\sum_{n=1}^{|\mathcal{S}_1|} K_n\,,$$

$$x_n = \frac{zK_n}{\sum_{n=1}^{|\mathcal{S}_1|} K_n}\,.$$

Hence the total delay consumed by the links in the set, $\mathcal{S}_1$, will be

$$\begin{aligned} \sum_{ij\in\mathcal{S}_1} D_{i,j}^{\mathrm{DT}} &= \sum_{n=1}^{|\mathcal{S}_1|} K_n(\rho L - \ln x_n) \\ &= \sum_{i,j\in\mathcal{S}_1} K_{i,j}\left(\rho L - \ln\left(\frac{zK_{i,j}}{\sum_{i,j\in\mathcal{S}_1} K_{i,j}}\right)\right), \end{aligned}$$

where the solution in (5.18) is obtained.

Similarly, the second optimization problem in (5.17) can be written as

$$\begin{aligned} \min \quad & \sum_{i,j\in\mathcal{S}_2} D_{i,j}^{\mathrm{CT}} \\ \text{s.t.} \quad & \sum_{i,j\in\mathcal{S}_2} \epsilon_{i,j}^{\mathrm{CT}} \le \epsilon - z\,. \end{aligned}$$

Again, define $x_n = \epsilon_{i,j}^{\mathrm{CT}}$ and

$$C_n = \frac{2}{\rho \ln\left(1 + \frac{\mathrm{SINR}_{i,j}^n + f(\mathrm{SINR}_{i,r}^n, \mathrm{SINR}_{r,j}^n)}{1+\rho}\right)},$$

where $n \in [1, \ldots, |\mathcal{S}_2|]$. Using the Kuhn-Tucker condition, its optimal solution can be found from

$$\begin{aligned} -\frac{C_1}{x_1} + \lambda &= 0\,, \\ &\vdots \\ -\frac{C_{|\mathcal{S}_2|}}{x_{|\mathcal{S}_2|}} + \lambda &= 0\,, \\ \lambda\left(\sum_{n=1}^{|\mathcal{S}_2|} x_n - \epsilon + z\right) &= 0\,. \end{aligned}$$

Hence the Lagrange multiple λ and the optimal solutions of x_n should be

$$\begin{aligned} \lambda &= \frac{1}{\epsilon - z}\sum_{n=1}^{|\mathcal{S}_2|} C_n\,, \\ x_n &= \frac{C_n(\epsilon - z)}{\left(\sum_{n=1}^{|\mathcal{S}_2|} C_n\right)}\,. \end{aligned}$$

Hence the overall delay by the links in the set, $\mathcal{S}_2$, is

$$\begin{aligned} \sum_{i,j\in\mathcal{S}_2} D_{i,j}^{\mathrm{CT}} &= \sum_{n=1}^{|\mathcal{S}_2|} C_n(\rho L - \log x_n) \\ &= \sum_{i,j\in\mathcal{S}_2} C_{i,j}\left(\rho L - \ln\left(\frac{C_{i,j}(\epsilon - z)}{\sum_{i,j\in\mathcal{S}_2} C_{i,j}}\right)\right)\,. \end{aligned}$$

A.4 PROOF OF THEOREM 5.2

Considering the network scenario that multi-node transmissions are enabled along the same route using the space-time reuse scheme, the minimum delay per hop using non-interference subtraction transmission is

$$D_{\text{NON}} = \frac{\rho L - \ln \epsilon_{ph}}{\rho \ln\left(1 + \frac{p}{(1+\rho)(N_0 + \sum_{i=1}^{\infty}(iK+1)^{-\alpha}p + \sum_{i=1}^{\infty}(iK-1)^{-\alpha}p)}\right)}$$

and the minimum delay per hop using interference subtraction is

$$D_{\text{IS}} = \frac{\rho L - \ln \epsilon_{ph}}{\rho \ln\left(1 + \frac{p}{(1+\rho)(N_0 + \sum_{i=1}^{\infty}(iK+1)^{-\alpha}p)}\right)}.$$

Assuming the system is in an interference-limited region, in which white noise power $N_0 \ll p$, the delay performance gain is

$$g = \frac{D_{\text{IS}}}{D_{\text{NON}}} = \frac{\ln\left(1 + \frac{1}{(1+\rho)(\sum_{i=1}^{\infty}(iK+1)^{-\alpha} + \sum_{i=1}^{\infty}(iK-1)^{-\alpha})}\right)}{\ln\left(1 + \frac{1}{(1+\rho)(\sum_{i=1}^{\infty}(iK+1)^{-\alpha})}\right)}. \tag{A.2}$$

A.4.1 Numerator of g

$$g_{\text{NUM}} = \ln\left(1 + \frac{1}{(1+\rho)\left(\underbrace{\sum_{i=1}^{\infty}(iK+1)^{-\alpha}}_{\mathbf{T}_1} + \underbrace{\sum_{i=1}^{\infty}(iK-1)^{-\alpha}}_{\mathbf{T}_2}\right)}\right).$$

We can obtain the bounds for $\mathbf{T}_1$ and $\mathbf{T}_2$ as follows:

$$\sum_{i=1}^{\infty}(iK+i)^{-\alpha} < \mathbf{T}_1 < \sum_{i=1}^{\infty}(iK)^{-\alpha}$$
$$\Rightarrow (1+K)^{-\alpha}\sum_{i=1}^{\infty} i^{-\alpha} < \mathbf{T}_1 < K^{-\alpha}\sum_{i=1}^{\infty} i^{-\alpha}.$$

Especially, we are interested in the result when the path-loss exponent is from 2 to 4. Since $\sum_{i=1}^{\infty} i^{-\alpha}$ is converged when $\alpha > 1$, we can directly get the result from zeta function [105], that is, zeta$[2] = \frac{\pi^2}{6}$, zeta$[3] = 1.202$, and zeta$[4] = \frac{\pi^4}{90}$. Therefore, $\mathbf{T}_1$ is bounded by two finite boundaries.

Similarly, we can obtain the bound for $\mathbf{T}_2$, which is

$$\begin{aligned}&\sum_{i=1}^{\infty}(iK)^{-\alpha} < \mathbf{T}_2 < \sum_{i=1}^{\infty}(iK-i)^{-\alpha}\\ &\Rightarrow K^{-\alpha}\text{zeta}[\alpha] < \mathbf{T}_2 < (K-1)^{-\alpha}\text{zeta}[\alpha]\,.\end{aligned}$$

Then the bound of $\mathbf{T}_1 + \mathbf{T}_2$ is

$$0 < (K^{-\alpha}+(K+1)^{-\alpha})\text{zeta}[\alpha] < \mathbf{T}_1+\mathbf{T}_2 < (K^{-\alpha}+(K-1)^{-\alpha})\text{zeta}[\alpha]\,.$$

Finally, the bound performance of g_{NUM} is

$$\begin{aligned}\ln\left(1+\frac{K^{\alpha}(K-1)^{\alpha}}{(1+\rho)(K^{\alpha}+(K-1)^{\alpha})\text{zeta}[\alpha]}\right) &< g_{\text{NUM}}\\ &< \ln\left(1+\frac{K^{\alpha}(K+1)^{\alpha}}{(1+\rho)(K^{\alpha}+(K+1)^{\alpha})\text{zeta}[\alpha]}\right).\end{aligned}$$

A.4.2 Denominator of *g*

$$g_{\text{DEN}} = \ln\left(1+\frac{1}{(1+\rho)\underbrace{\left(\sum_{i=1}^{\infty}(iK+1)^{-\alpha}\right)}_{\mathbf{T}_3}}\right).$$

We can derive the following:

$$\begin{aligned}&\sum_{i=1}^{\infty}(iK+i)^{-\alpha} < \mathbf{T}_3 < \sum_{i=1}^{\infty}(iK)^{-\alpha}\\ &\Rightarrow 0 < (1+K)^{-\alpha}\text{zeta}[\alpha] < \mathbf{T}_3 < K^{-\alpha}\text{zeta}[\alpha]\,.\end{aligned}$$

Hence the bound performance of g_{DEN} is

$$\ln\left(1+\frac{K^{\alpha}}{(1+\rho)\text{zeta}[\alpha]}\right) < g_{\text{DEN}} < \ln\left(1+\frac{(K+1)^{\alpha}}{(1+\rho)\text{zeta}[\alpha]}\right).$$

Using the two bounds in (A.2) leads to the result.

A.5 PROOF OF EQUATION (6.26)

According to Figure A.1, keeping r as a constant and moving θ to $\frac{\pi}{2}$, we have the expectation

$$\begin{aligned}\mathrm{E}[\beta] &> \frac{\sqrt{2}}{d_{s,d}^{\frac{\alpha}{2}} K}\mathrm{E}\left[\left(\frac{1}{4}+\frac{r^2}{d_{s,d}^2}\right)^{\frac{\alpha}{4}}\right] \\ &= \frac{2\sqrt{2}\lambda\pi}{d_{s,d}^{\frac{\alpha}{2}} K}\int_0^{\infty}\left(\frac{1}{4}+\frac{r^2}{d_{s,d}^2}\right)^{\frac{\alpha}{4}} re^{-\lambda\pi r^2}dr\,.\end{aligned} \tag{A.3}$$

Let $y = d_{s,d}^2/4 + r^2$, $2rdr = dy$, and $r^2 = y - d_{s,d}^2/4$; (A.3) equals to

$$\mathrm{E}[\beta] > \frac{\sqrt{2}\lambda\pi e^{\frac{\lambda\pi d_{s,d}^2}{4}}}{d_{s,d}^{\frac{\alpha}{2}} K}\int_{\frac{d_{s,d}^2}{4}}^{\infty} y^{\frac{\alpha}{4}} e^{-\lambda\pi y} dy\,. \tag{A.4}$$

Further let $\lambda\pi y = t$ and $y = \frac{t}{\lambda\pi}$; (A.4) equals to

$$\begin{aligned}\mathrm{E}[\beta] &\geq \frac{\sqrt{2}\lambda\pi e^{\frac{\lambda\pi d_{s,d}^2}{4}}}{d_{s,d}^{\frac{\alpha}{2}} K(\lambda\pi)^{\frac{\alpha+4}{4}}}\int_{\frac{\lambda\pi d_{s,d}^2}{4}}^{\infty} t^{\frac{\alpha+4}{4}-1} e^{-t} dt \\ &= \frac{\sqrt{2}e^{\frac{\lambda\pi d_{s,d}^2}{4}}}{d_{s,d}^{\frac{\alpha}{2}} K(\lambda\pi)^{\frac{\alpha}{4}}}\Gamma\left(\frac{\alpha+4}{4}, \frac{\lambda\pi d_{s,d}^2}{4}\right).\end{aligned}$$

A.6 PROOF OF THEOREM 7.1

According to the Kuhn-Tucker condition (p. 244: KKT conditions for convex problems [69]), the inequality constraint in (7.5) can be converted to the equality constraint and have the target outage probability

$$\frac{1}{2}d_{s,d}^{\alpha}\left(d_{s,r}^{\alpha}+\frac{p}{q}d_{r,d}^{\alpha}\right)\frac{(2^{2b}-1)^2}{p^2} = \eta\,.$$

Then we obtain

$$q = f(p) = \frac{Ap}{p^2 - B}, \tag{A.5}$$

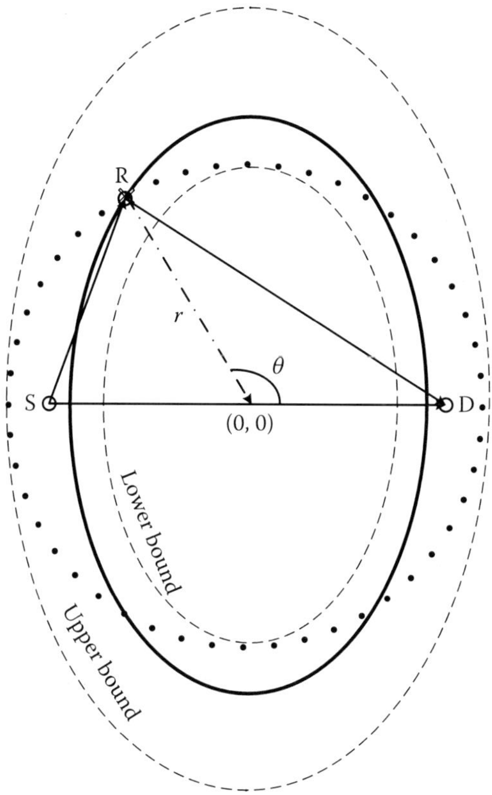

FIGURE A.1 Geometric bounds for E[β].

where $A = \mu d_{s,d}^{\alpha} d_{r,d}^{\alpha}/2\eta$, $B = \mu d_{s,d}^{\alpha} d_{s,r}^{\alpha}/2\eta$, $\mu = (2^{2b}-1)^2$, and η is the outage constraint.

Substituting (A.5) into $p+q$, and minimizing with regard to p, we have the solution

$$p^2 = \frac{A+2B}{2} \pm \frac{\sqrt{A^2+8AB}}{2}.$$

To be a valid solution for q, the solution must satisfy $p^2 > B$ in (A.5). So, we have a unique solution given by

$$p^* = \sqrt{\frac{A+2B}{2} + \frac{\sqrt{A^2+8AB}}{2}}.$$

Using this result in (A.5) leads to (7.6).

A.7 PROOF OF THEOREM 7.6

Figure A.2 illustrates the integration method of $\mathrm{E}[\beta]$. The integration is performed from the best relay location at the bottom with the minimum value $\frac{1}{2K}$ to infinity; r is the selected relay distance to the destination. Since the minimum β at point a is smaller than the minimum β'[3] at point b, when β increases to point b, both β and β' have the same power ratio on the same cut. It is worth noting that the two circles on the same cut are the set of relay locations that achieve the same power ratio for the conventional cooperation and the optimal cooperation, respectively. Therefore, we have the expected power ratio

$$\begin{aligned}
\mathrm{E}[\beta] &= \int_0^{\infty} \beta f(r)dr > \int_0^{\frac{d_{s,d}}{2}} \beta f(r)dr + \int_{\frac{d_{s,d}}{2}}^{\infty} \beta' f(r)dr \\
&> \int_0^{\frac{d_{s,d}}{2}} \frac{1}{2K} f(r)dr + \frac{\sqrt{2}}{K}\int_{\frac{d_{s,d}}{2}}^{\infty} \sqrt{\frac{1}{4} + \frac{(r-\frac{d_{s,d}}{2})^2}{d_{s,d}^2}}\, f(r)dr \\
&= \frac{1-e^{-\rho}}{2K} + \frac{\sqrt{2}}{K}\int_{\frac{d_{s,d}}{2}}^{\infty} \sqrt{\frac{1}{4} + \frac{(r-\frac{d_{s,d}}{2})^2}{d_{s,d}^2}}\, f(r)dr.
\end{aligned}$$

[3]Note that β' is the average power ratio of the conventional DAF cooperation scheme, which can be found in (6.1).

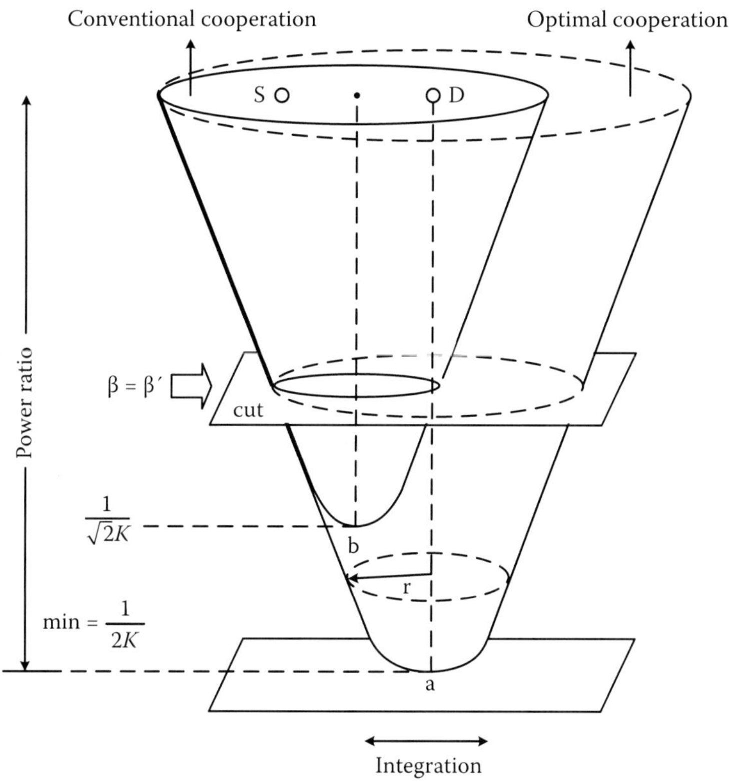

FIGURE A.2 Integration from geometric point of view.

Bibliography

[1] J. N. Laneman, D. N. Tse, and G. W. Wornell. Cooperative diversity in wireless networks: Efficient protocols and outage behavior. *IEEE Trans. Info. Theory*, 50(12):3062–3080, Dec. 2004.

[2] A. Scaglione, D. Goeckel, and J. N. Laneman. Cooperative communications in mobile ad-hoc networks: Rethinking the link abstraction. *IEEE Sig. Proc. Mag.*, 23:18–29, 2006.

[3] Zhengguo Sheng, Zhiguo Ding, and K. K. Leung. Distributed and power efficient routing in wireless cooperative networks. In *Proc. IEEE International Conference on Communications (ICC)*, pages 1–5, 2009.

[4] A. Ozgur, O. Leveque, and D. N. C. Tse. Hierarchical cooperation achieves optimal capacity scaling in ad hoc networks. *IEEE Trans. Inf. Theory*, 53(10):3549–3572, Oct. 2007.

[5] Chiung-Jang Chen and Li-Chun Wang. Enhancing coverage and capacity for multiuser MIMO system by utilizing scheduling. *IEEE Trans. Wireless Comm.*, 5:1148–1157, May 2006.

[6] T. Yoo and A. Goldsmith. On the optimality of multiantenna broadcast scheduling using zero-forcing beamforming. *IEEE J. Select. Areas in Comm.*, 24:528–540, Mar. 2006.

[7] V. Srivastava and M. Motani. Cross-layer design: A survey and the road ahead. *IEEE Comm. Mag.*, 43(12):112–119, Dec. 2005.

[8] Yinman Lee, Ming hung Tsai, and Sok-Ian Sou. Performance of decode and forward cooperative communications with multiple dual-hop relays over nakagami-m fading channels. *IEEE Trans. Wireless Comm.*, 8(6):2853–2859, 2009.

[9] V. Asghari and S. Aissa. End-to-end performance of cooperative relaying in spectrum-sharing systems with quality of service requirements. *IEEE Trans. Veh. Tech.*, 60(6):2656–2668, 2011.

[10] Meixia Tao and Yuan Liu. A network flow approach to throughput maximization in cooperative OFDMA networks. *IEEE Trans. Wireless Comm.*, 12(3):1138–1148, 2013.

[11] M. Elhawary and Z. J. Haas. Energy-efficient protocol for cooperative networks. *IEEE/ACM Trans. Networking*, 19(2):561–574, 2011.

[12] Mao Xu, Jun Zhang, and Weidong Wang. Delay fairness aware ARQ protocol for two-user cooperative networks. In *2nd International Conference on Computer Engineering and Technology (ICCET)*, volume 2, pages V2–234–V2–238, Apr. 2010.

[13] Sanquan Song, D. L. Goeckel, and D. Towsley. Collaboration improves the connectivity of wireless networks. In *25th IEEE International Conference on Computer Communications (INFOCOM)*, pages 1–11, 2006.

[14] R. Babaee and N. C. Beaulieu. Cross-layer design for multihop wireless relaying networks. *IEEE Trans. Wireless Comm.*, 9(11):3522–3531, 2010.

[15] S. S. Ikki and M. H. Ahmed. Performance analysis of adaptive decode-and-forward cooperative diversity networks with best-relay selection. *IEEE Trans. Comm.*, 58(1):68–72, 2010.

[16] Sung-Rae Cho, Wan Choi, and Kaibin Huang. QoS provisioning relay selection in random relay networks. *IEEE Trans. Veh. Tech.*, 60(6):2680–2689, 2011.

[17] John R. Barry, Edward A. Lee, and David G. Messerschmitt. *Digital Communication*. Kluwer Academic Publishers, Dordrecht, 3rd edition, 2003.

[18] Z. Ding, T. Ratnarajah, and C. Cowan. On the multiplexing-diversity tradeoff for wireless superposition cooperative multiple access systems. *IEEE Trans. Signal Process.*, 55(9):4627–4638, 2007.

[19] T. H. Clausen, G. Hansen, L. Christensen, and G. Behrmann. The optimized link state routing protocol evaluation through experiments and simulation. In *Proc. IEEE Symposium on "Wireless Personal Mobile Communications"*, 2001. Available at http://hipercom.inria.fr/olsr/#papers.

[20] S. Zhu and K. K. Leung. Distributed cooperative routing for UWB ad-hoc networks. In *Proc. IEEE International Conference on Communication, ICC*, pages 3339–3344, June 2007.

[21] Charles E. Perkins and Pravin Bhagwat. Highly dynamic destination-sequenced distance-vector routing (DSDV) for mobile computers. *SIGCOMM Comput. Comm. Rev. (Proc. of ACM SIGCOMM)*, 24(4):234–244, 1994.

[22] A. E. Khandani. *Cooperative Routing in Wireless Networks*. PhD thesis, MIT, 2004.

[23] F. Li, K. Wu, and A. Lippman. Energy-efficient cooperative routing in multi-hop wireless ad hoc networks. In *Proc. IEEE IPCCC*, pages 222–230, 2006.

[24] C. Pandana, W. P. Siriwongpairat, T. Himsoon, and K. J. R. Liu. Distributed cooperative routing algorithm for maximizing network lifetime. In *IEEE Wireless Communications and Networking Conference*, pages 451–456, 2006.

[25] Weiyan Ge, Junshan Zhang, and Guoliang Xue. Joint clustering and optimal cooperative routing in wireless sensor networks. In *Proc. IEEE International Conference on Communication (ICC)*, pages 2216–2220, 2008.

[26] Ahmed S. Ibrahim, Zhu Han, and K. J. Ray Liu. Distributed energy-efficient cooperative routing in wireless networks. In *Proc. IEEE Global Telecommunication Conference, Globecom*, pages 4413–4418, Nov. 2007.

[27] S. Ghez, S. Verdu, and D. C. Schwartz. Stability properties of slotted aloha with multipacket reception capability. *IEEE Trans. Auto. Cont.*, 33:640–649, 1988.

[28] M. Howlader, K. Mostofa, and B. D. Woerner. System architecture for implementing multiuser detector within an ad-hoc network. In *Proc. IEEE Military Communication Conference (MILCOM)*, volume 2, pages 1119–1123, 2001.

[29] C. Comaniciu, N. B. Mandayam, and H. V. Poor. *Wireless Networks: Multiuser Detection in Cross-Layer Design*. Springer, New York, 2005.

[30] Surendra Boppana and John M. Shea. Overlapped transmission in wireless ad hoc networks. In *Proc. International Conference on Comuunication, Circuits and Systems*, volume 2, pages 1309–1314, June 2006.

[31] I. E. Telatar and R. G. Gallager. Combining queueing theory with information theory for multiaccess. *IEEE J. Sel. Areas Comm.*, 13:963–969, Aug. 1995.

[32] J. Proakis. *Digital Communication*. McGraw-Hill, New York, 2000.

[33] R. G. Gallager. *Information Theory and Reliable Communication*. John Wiley and Sons, Inc., New York, 1968.

[34] Ning Wen and Randall A. Berry. Reliability constrained packet-sizing for linear multi-hop wireless networks. In *Proc. IEEE International Symposium on Information Theory*, pages 16–20, July 2008.

[35] Gerhard Kramer, Ivana Maric, and Roy D. Yates. *Cooperative Communications*. Now Publishers Inc., Hanover, MA, 2006.

[36] Zhihang Yi and Il-Min Kim. Diversity order analysis of the decode-and-forward cooperative networks with relay selection. *IEEE Trans. Wireless Comm.*, 7(5):1792–1799, May 2008.

[37] J. Luo, R. S. Blum, L. J. Cimini, L. J. Greenstein, and A. M. Haimovich. Decode-and-forward cooperative diversity with power allocation in wireless networks. *IEEE Trans. Wireless Comm.*, 6(3):793–799, Mar. 2007.

[38] Jialing Zheng and Maode Ma. QoS-aware cooperative medium access control for MIMO ad-hoc networks. *IEEE Comm. Letters*, 14(1):48–50, Jan. 2010.

[39] A. S. Ibrahim and K. J. R. Liu. Mitigating channel estimation error with timing synchronization tradeoff in cooperative communications. *IEEE Trans. Signal Processing*, 58(1):337–348, Jan. 2010.

[40] M. Stojanovic and J. Preisig. Underwater acoustic communication channels: Propagation models and statistical characterization. *IEEE Comm. Mag.*, 47(1):84–89, Jan. 2009.

[41] H. Adam, C. Bettstetter, and S. M. Senouci. Adaptive relay selection in cooperative wireless networks. In *Proc. IEEE Personal, Indoor and Mobile Radio Communications (PIMRC)*, pages 1–5, 2008.

[42] X.-L. Huang, G. Wang, and F. Hu. Multitask spectrum sensing in cognitive radio networks via spatiotemporal data mining. *IEEE Trans. Veh. Tech.*, 62(2):809–823, 2013.

[43] B. Seshasayee and K. Schwan. Energy-effcient device scheduling through contextual timeouts. Georgia Institute of Technology, Technical Report, 2005.

[44] W. Ye, J. Heidemann, and D. Estrin. An energy-efficient mac protocol for wireless sensor networks. In *Proc. INFOCOM 2002. The 21st Joint Conference of the IEEE Computer and Communications Societies. IEEE*, volume 3, pages 1567–1576, 2002.

[45] J.-H. Chang and L. Tassiulas. Energy conserving routing in wireless ad-hoc networks. In *Proc. INFOCOM 2000. The 9th Joint Conference of the IEEE Computer and Communications Societies. IEEE*, pages 22–31, 2000.

[46] Y. Jian, E. Liu, Y. Wang, and et al. Scale-free model for wireless sensor networks. In *Proc. IEEE WCNC*, pages 2329–2332, 2012.

[47] Y. Wang, E. Liu, X. Zheng, and et al. Energy-aware complex network model with compensation. In *Proc. IEEE WiMob*, pages 472–476, 2013.

[48] J. E. Wieselthier, G. D. Nguyen, and A. Ephremides. Algorithms for energy-efficient multicasting in static ad hoc wireless networks. *ACM Mobile Networks and Applications (MONET)*, 6(3):251–263, 2001.

[49] J. N. Laneman, D. N. C. Tse, and G. W. Wornell. Cooperative diversity in wireless networks: Efficient protocols and outage behavior. *IEEE Trans. Info. Theory*, 50(12):3062–3080, 2004.

[50] A. Sendonaris, E. Erkip, and B. Aazhang. User cooperation diversity—Part I: System description. *IEEE Trans. Comm.*, 51(11):1927–1938, 2003.

[51] A. Bletsas, A. Khisti, D. P. Reed, and A. Lippman. A simple cooperative diversity method based on network path selection. *IEEE J. Sel. Areas Comm.*, 24(3):659–672, 2006.

[52] E. Liu, Q. Zhang, and K. Leung. Relay-assisted transmission with fairness constraint for cellular networks. *IEEE Trans. Mobile Computing*, 11(2):230–239, 2012.

[53] C. Wang, H. Farhadi, and M. Skoglund. Achieving the degrees of freedom of wireless multi-user relay networks. *IEEE Trans. Comm.*, 60(9):2612–2622, 2012.

[54] A. E. Khandani, J. Abounadi, E. Modiano, and L. Zheng. Cooperative routing in static wireless networks. *IEEE Trans. Comm.*, 55(11):2185–2192, 2007.

[55] Y.-S. Tu and G. J. Pottie. Coherent cooperative transmission from multiple adjacent antennas to a distant stationary antenna through AWGN channels. In *Proc. IEEE 55th Vehicular Technology Conference VTC Spring 2002*, pages 130–134, 2002.

[56] R. Ahlswede, Ning Cai, S.-Y. R. Li, and R. W. Yeung. Network information flow. *IEEE Trans. Info. Theory*, 46(4):1204–1216, July 2000.

[57] R. F. Wyrembelski, T. J. Oechtering, and H. Boche. Decode-and-forward strategies for bidirectional relaying. In *IEEE PIMRC '08*, pages 1–6, Sep. 15–18, 2008.

[58] S. Katti, S. Gollakota, and D. Katabi. Embracing wireless interference: analog network coding. In *SIGCOMM '07*, pages 397–408, Aug. 27–31, 2007.

[59] S. Zhang, S. Liew, and P. Lam. Hot topic: Physical layer network coding. In *ACM MobiCom '06*, pages 358–365, Sep. 23–26, 2006.

[60] Ying-Chang Liang and Rui Zhang. Optimal analogue relaying with multi-antennas for physical layer network coding. In *IEEE ICC '08*, pages 3893–3897, May 19–23, 2008.

[61] Namyoon Lee, Hyun Jong Yang, and Joohwan Chun. Achievable sum-rate maximizing AF relay beamforming scheme in two-way relay channels. In *IEEE ICC Workshops '08*, pages 300–305, May 19–23, 2008.

[62] Chunguo Li, Luxi Yang, and Wei-Ping Zhu. Two-way MIMO relay precoder design with channel state information. *IEEE Trans. Comm.*, 56(12):3358–3363, Dec. 2010.

[63] Chee Yen Leow, Zhiguo Ding, Kin K. Leung, and Dennis L. Goeckel. On the study of analogue network coding for multi-pair, bidirectional relay channels. *IEEE Trans. Wireless Comm.*, 10(2):670–681, Feb. 2011.

[64] Kyoung-Jae Lee, Hakjea Sung, Eunsung Park, and Inkyu Lee. Joint optimization for one and two-way MIMO AF multiple-relay systems. *IEEE Trans. Wireless Comm.*, 9(12):3671–3681, Dec. 2010.

[65] Shengyang Xu and Yingbo Hua. Optimal design of spatial source-and-relay matrices for a non-regenerative two-way MIMO relay system. *IEEE Trans. Wireless Comm.*, 10(5):1645–1655, Mar. 2011.

[66] Ezio Biglieri, Robert Calderbank, Anthony Constantinides, Andrea Goldsmith, Arogyaswami Paulraj, and H. Vincent Poor. *MIMO Wireless Communications*. Cambridge University Press, Cambridge, UK, 2007.

[67] Gilbert Strang. *Linear Algebra and Its Applications*. Thomson Learning, New York, 3rd edition, 1988.

[68] John M. Cioffi, Glen P. Dudevoir, M. Vedat Eyuboglu, and G. David Forney. MMSE decision-feedback equalizers and coding. II. Coding results. *IEEE Trans. Communications*, 43(10):2595–2604, Oct. 1995.

[69] S. Boyd and L. Vandenberghe. *Convex Optimization*. Cambridge University Press, Cambridge, UK, 2003.

[70] Ingmar Hammerstrm and Armin Wittneben. Power allocation schemes for amplify-and-forward MIMO-OFDM relay links. *IEEE Trans. Wireless Comm.*, 6(8):2798–2802, Aug. 2007.

[71] M. Grant and S. Boyd. CVX: Matlab software for disciplined convex programming, version 1.21. http://cvxr.com/cvx, Apr. 2011.

[72] Michael Grant and S. Boyd. Graph implementations for nonsmooth convex programs. In V. Blondel, S. Boyd, and H. Kimura, editors, *Recent Advances in Learning and Control*, Lecture Notes in Control and Information Sciences, pages 95–110. Springer-Verlag, Berlin, 2008.

[73] A. Nosratinia, T. E. Hunter, and A. Hedayat. Cooperative communication in wireless networks. *IEEE Communications Magazine*, 42(10):74–80, Oct. 2004.

[74] W. Zhuang and M. Ismail. Cooperation in wireless communication networks. *IEEE Wireless Comm. Mag.*, 19(2):10–20, Apr. 2012.

[75] A. Ikhlef, D. S. Michalopoulos, and R. Schober. Max-max relay selection for relays with buffers. *IEEE Trans. Wireless Comm.*, 11(3):1124–1135, Mar. 2012.

[76] Zhengguo Sheng, K. K. Leung, and Zhiguo Ding. Cooperative wireless networks: From radio to network protocol designs. *IEEE Comm. Mag.*, 49(5):64–69, May 2011.

[77] Yifei Wei, F. R. Yu, and Mei Song. Distributed optimal relay selection in wireless cooperative networks with finite-state markov channels. *IEEE Trans. Veh. Tech.*, 59(5):2149–2158, June 2010.

[78] Zhengguo Sheng, Bong Jun Ko, and K. K. Leung. Power efficient decode-and-forward cooperative relaying. *IEEE Wireless Comm. Letters*, 1(5):444–447, 2012.

[79] Erwu Liu, Qinqing Zhang, and K. K. Leung. Residual energy-aware cooperative transmission (react) in wireless networks. In *IEEE WOCC'10*, pages 1–6, 2010.

[80] Dan Chen, Hong Ji, Xi Li, and Kun Zhao. A novel multi-relay selection and power allocation optimization scheme in cooperative networks. In *IEEE WCNC'10*, pages 1–6, 2010.

[81] K. Vardhe, D. Reynolds, and B. D. Woerner. Joint power allocation and relay selection for multiuser cooperative communication. *IEEE Trans. Wireless Comm.*, 9(4):1255–1260, 2010.

[82] Erwu Liu, Qinqing Zhang, and K. K. Leung. Connectivity in selfish, cooperative networks. *IEEE Comm. Letters*, 14(10):936–938, Oct. 2010.

[83] J. N. Laneman and G. W. Wornell. Distributed space-time-coded protocols for exploiting cooperative diversity in wireless networks. *IEEE Trans. Info. Theory*, 49:2415–2425, Oct. 2003.

[84] J. N. Laneman. Limiting analysis of outage probabilities for diversity schemes in fading channels. In *IEEE GLOBECOM'03*, pages 1242–1246, 2003.

[85] Rafael Asorey Cacheda, Daniel Castro García, Antonio Cuevas, Francisco Javier González Castaño, and et al. QoS requirements for multimedia services. In Giovanni Giambene, editor, *Resource Management in Satellite Networks*, pages 67–94. Springer, New York, 2007.

[86] G. A. Shah, Weifa Liang, and O. B. Akan. Cross-layer framework for QoS support in wireless multimedia sensor networks. *IEEE Trans. Multimedia*, 14(5):1442–1455, 2012.

[87] Quang Trung Duong. *On Cooperative Communications and Its Applications to Mobile Multimedia.* Blekinge Institute of Technology, Karlskrona, Sweden, 2010.

[88] R. Trestian, O. Ormond, and G.-M. Muntean. Energy-quality-cost tradeoff in a multimedia-based heterogeneous wireless network environment. *IEEE Trans. Broadcasting*, 59(2):340–357, 2013.

[89] G. Foschini and M. Gans. On limits of wireless communication in a fading environment when using multiple antennas. *Wireless Personal Comm.*, 6(3):311–335, Mar. 1998.

[90] Emre Telatar. Capacity of multi-antenna gaussian channels. *Euro. Trans. Telecom.*, 10:585–595, Nov./Dec. 1999.

[91] Cheng-Xiang Wang, Xuemin Hong, Xiaohu Ge, Xiang Cheng, Gong Zhang, and J. Thompson. Cooperative MIMO channel models: A survey. *IEEE Comm. Mag.*, 48(2):80–87, 2010.

[92] G. G. de Oliveira Brante, M. T. Kakitani, and R. Demo Souza. Energy efficiency analysis of some cooperative and non-cooperative transmission schemes in wireless sensor networks. *IEEE Trans. Comm.*, 59(10):2671–2677, 2011.

[93] Qinghua Shi and Y. Karasawa. Error probability of opportunistic decode-and-forward relaying in nakagami-m fading channels with arbitrary m. *IEEE Wireless Comm. Letters*, 2(1):86–89, 2013.

[94] Zhengguo Sheng, Bong Jun Ko, A. Swami, Kang-Won Lee, and K. K. Leung. Power efficiency of decode-and-forward cooperative relaying. In *The Military Communications Conference (MILCOM)*, pages 543–548, 2010.

[95] Zhiguo Ding and K. K. Leung. Cross-layer routing using cooperative transmission in vehicular ad-hoc networks. *IEEE J. Select. Areas Comm.*, 29(3):571–581, 2011.

[96] T. Winter and P. Thubert. *RPL: IPv6 routing protocol for low power and lossy networks.* Internet Engineering Task Force (IETF), Draft, 2012.

[97] Gyorgy Szabo and Gabor Fath. Evolutionary games on graphs. *Physics Reports*, 2007.

[98] Youngchul Sung, Saswat Misra, Lang Tong, and Anthony Ephremides. Cooperative routing for distributed detection in large sensor networks. *IEEE J. Sel. Areas Comm.*, 25(2):471–483, Feb. 2007.

[99] M. Z. Siam, M. Krunz, and O. Younis. Energy-efficient clustering/routing for cooperative MIMO operation in sensor networks. In *IEEE INFOCOM*, pages 621–629, Apr. 2009.

[100] Gentian Jakllari, Srikanth V. Krishnamurthy, Michalis Faloutsos, Prashant V. Krishnamurthy, and Ozgur Ercetin. A cross-layer framework for exploiting virtual MISO links in mobile ad hoc networks. *IEEE Trans. Mobile Computing*, 6(6):579–594, June 2007.

[101] R. Madan, N. B. Mehta, A. F. Molisch, and Jin Zhang. Energy-efficient decentralized cooperative routing in wireless networks. *IEEE Trans. Auto. Cont.*, 54(3):512–527, Mar. 2009.

[102] Mung Chiang, S. H. Low, A. R. Calderbank, and J. C. Doyle. Layering as optimization decomposition: A mathematical theory of network architectures. *Proc. IEEE*, 95(1):255–312, Jan. 2007.

[103] Loc Bui, A. Eryilmaz, R. Srikant, and Xinzhou Wu. Asynchronous congestion control in multi-hop wireless networks with maximal matching-based scheduling. *IEEE/ACM Trans. Networking*, 16(4):826–839, Aug. 2008.

[104] G. Sharma, R. R. Mazumdar, and N. B. Shroff. The complexity of scheduling in wireless networks. In *Proc. ACM MOBICOM*, pages 227–238, 2006.

[105] Edward Charles Titchmarsh and D. R. Heath-Brown. *The Theory of the Riemann Zeta-function*. Oxford University Press, New York, 1987.

Index

9780081027523

9780081027523